Pseudo-differential Operators and the Nash–Moser Theorem

Pseudo-differential Operators and the Nash–Moser Theorem

Serge Alinhac
Patrick Gérard

Translated by Stephen S. Wilson

Graduate Studies in Mathematics
Volume 82

American Mathematical Society
Providence, Rhode Island

This work was originally published in French by EDP Science, Paris, under the title "Opérateurs pseudo-différentiels et théorème de Nash-Moser" © 1991 InterEditions and Editions de CNRS. The present translation was created under license for the American Mathematical Society and is published by permission.

Translated by Stephen S. Wilson

2000 *Mathematics Subject Classification*. Primary 35–02;
Secondary 35Sxx, 47G30, 47N20.

For additional information and updates on this book, visit
www.ams.org/bookpages/gsm-82

Library of Congress Cataloging-in-Publication Data

Alinhac, S. (Serge)
[Opérateurs pseudo-différentiels et théorème de Nash-Moser. English]
Pseudo-differential operators and the Nash-Moser theorem / Serge Alinhac, Patrick Gérard ; translated by Stephen S. Wilson.
p. cm.
Includes bibliographical references and index.
ISBN 978-0-8218-3454-1 (alk. paper)
1. Pseudodifferential operators. 2. Implicit functions. I. Gérard, Patrick, 1961– II. Title.

QA329.7.A4513 2007
515′.7242—dc22 2006047985

∞ The paper used in this book is acid-free and falls within the guidelines
established to ensure permanence and durability.
Visit the AMS home page at http://www.ams.org/

10 9 8 7 6 5 4 3 2 1 12 11 10 09 08 07

Contents

Preface to the English edition

We are happy to welcome the English translation of our book, which originally appeared under the title 'Opérateurs pseudo-différentiels et théorème de Nash–Moser' in 1991 (InterEditions/Editions du CNRS, Paris).

Though the world of partial differential equations has changed a lot during these years, we think that the elementary presentation of the subjects touched upon in our book is still up to date and can be useful; thus, we made no changes, except for correcting some misprints. On the other hand, several remarkable books on partial differential equations have appeared since: though their scopes largely exceed that of our book, we thought it relevant to mention them in our bibliography.

Finally, we wish to thank the translator, Dr. Stephen S. Wilson, and the editorial board of the AMS, who worked to produce this new edition of our work.

Orsay, November 2006 Serge Alinhac and Patrick Gérard

Translator's note

The numbering system I have used in my translation is essentially that employed by the authors in the original French edition so that the actual equation numbers etc. are the same in both versions. I did, however, make certain changes to the cross-referencing system: for example, to remove ambiguity, outside of Chapter II Exercise A.1 of that chapter may be referred to here as Exercise II.A.1 although within Chapter II it is referred to as Exercise A.1.

Cheltenham, February 2007 Stephen S. Wilson

General introduction

This book provides an elementary exposition intended for students who have completed four years' study of university level mathematics. A knowledge of the elements of functional analysis, Fourier analysis and distribution theory (including, in particular, Fourier analysis in $\mathcal{S}$ and $\mathcal{S}'$) is assumed. Chapter 0 contains a reminder of the notation, concepts and main results used in the remainder of the book (with references). On the other hand, no knowledge of partial differential equations is needed, although it will be beneficial to have received an initiation to the topic.

The book stems from a course on 'Pseudo-differential operators and the Nash–Moser theorem', presented at the Ecole Normale Supérieure (ENS) from October 1986 onwards, to second-year students studying for the degree of Master of Fundamental and Applied Mathematics and Computer Science.

Although the topics covered largely form the subject of research literature, we have striven to avoid any scholarly discussions, 'veiled references' and sibylline allusions, which might open chasms beneath the reader's footsteps. A particular presentation of the subject is selected and developed in each chapter: the commentary at the end of each chapter indicates the sources, differing approaches, certain current extensions, and the connections between the topics handled.

Finally, we have assembled numerous exercises, divided into two classes. Elementary exercises are intended to help readers assimilate the course and monitor their progress. Other more complex exercises, marked with an asterisk (∗), present recent developments which have sometimes only been published in journal articles: we crave their authors' forgiveness for this simplification! These exercises, unlike those in certain famous treatises, can be effectively solved by real students, as experience of the teaching at ENS has shown.

It was our wish that this text should also be useful to researchers as a simple and *self-contained* introduction to subjects with which they are unfamiliar.

The dual purpose of these notes led us to keep them short, sometimes at the expense of a certain denseness of the text (which we believe is essentially accessible to motivated students). In particular, we had in mind our many colleagues in 'applied mathematics' who wish to use the Nash–Moser theorem in their research or to keep themselves up to date on microlocal analysis, without delving into the arcana of the specialist literature: they will be able to read the desired chapters independently of each other.

The choice of the material presented is a matter of personal taste and of the fields of research of the authors who, incidentally, believe that certain difficult (nonlinear) problems cannot be solved without a sufficient knowledge of pseudo-differential operators.

The authors are indebted to numerous mathematicians (cited in the commentaries) who have inspired them to present the subjects dealt with, and, in particular, to L. Hörmander, to whom the mathematical contents of Chapter I and Section III.C are largely due. The Bibliography at the end of the book indicates the sources used.

While presenting important concepts which are the true starting points for numerous recent developments, we have sought to end up with real theorems: microlocal elliptic regularity; propagation of singularities; existence of solutions of quasilinear hyperbolic systems; existence of isometric embeddings; the Nash–Moser theorem. The plan of the book is as follows.

In Chapter I we present the 'minimal' theory of pseudo-differential operators, in a global context (on $\mathbb{R}^n$), which turns out to be very nice in practice. The main points here are the notion of the symbol, the symbolic calculus for operators, the action in Sobolev spaces and the invariance under change of coordinates. The text presents only a few concrete applications and the most technical proofs are brought together in the appendix, in order to enable the reader to obtain an overall view of the subject. The exercises in Chapter I, which are particularly numerous, provide an introduction to a number of variants of the theory proposed and present several applications notably to the analysis on compact manifolds.

Chapter II brings together three themes. Section A presents the Littlewood–Paley theory of 'dyadic decomposition' of distributions: this systematizes the natural division of the space of frequencies ξ according to their size $|\xi|$, associated with the classical symbolic calculus of Chapter I. This very simple theory allows one to rapidly obtain interesting properties of composite functions in Sobolev and Hölder spaces. Section B presents the concept of the wave front set and its links with pseudo-differential operators: this time it is a matter of the conical division of the space of frequencies ξ according to their directions $\xi \in S^{n-1}$, associated with the classical symbol homogeneities. Finally, Section C deals with hyperbolic energy inequalities for which pseudo-differential operators turn out to be an effective tool. Thus, Chapter II serves to present very useful applications of the 'dry theory' of Chapter I, while preparing the material and the concepts which will be needed in Chapter III.

The final chapter discusses certain problems of a nonlinear nature which arise in geometry or in analysis and which may be reduced to perturbation problems. The plan of this chapter reflects the various situations which one may encounter: 'elliptic' situations in which the usual Banach implicit function theorem suffices; 'fixed-point' situations, such as one often finds in nonlinear hyperbolic problems or again in the isometric embedding problem; and, finally, situations where the 'loss of derivatives' is too great and a Nash–Moser technique has to be used. The Nash–Moser theorem relies completely on the acquisition of *a priori* 'tame' inequalities; the reader who is already familiar with *a priori* inequalities (presented in Chapter I and Section III.C) will grasp the concept of 'tame' estimates through its clear link with Littlewood–Paley theory and the paradifferential calculus of J.-M. Bony (Section II.A).

This establishes the underlying cohesion of this book, which can be schematized in the accompanying diagram.

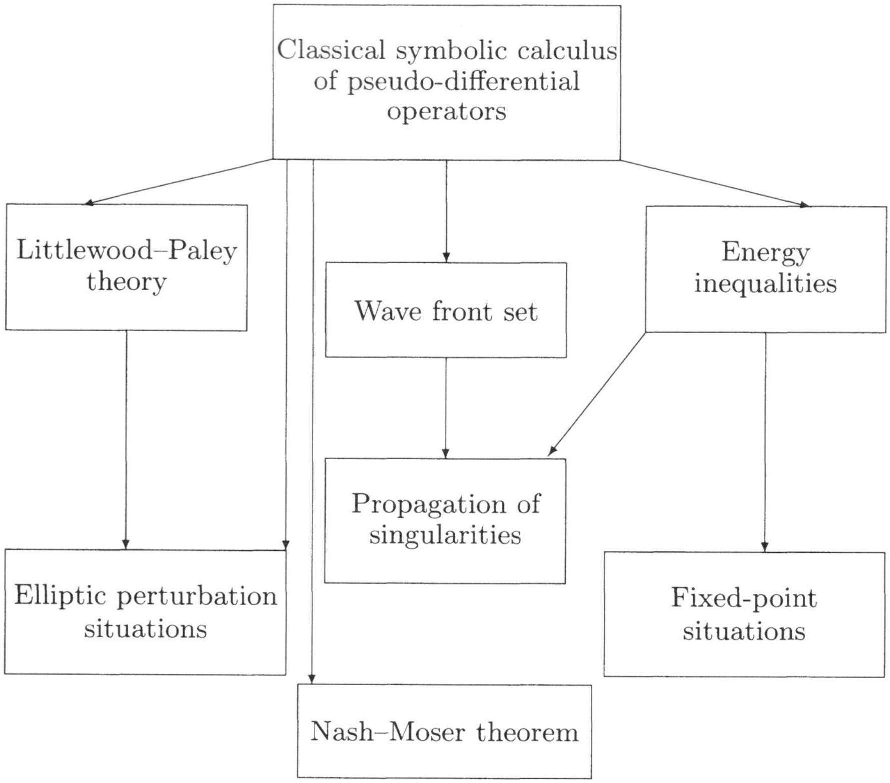

In this spirit, we were recently very happy to learn of the work of L. Hörmander [H9], explaining the links between pseudo-differential and paradifferential operators, fixed-point methods and the Nash–Moser theorem.

Finally, we are grateful to G. Ben Arous and J.B. Bost for their kind and valuable suggestions.

Notation and review of distribution theory

In this chapter, we introduce the various notation used in the book, while recalling a number of elements of distribution theory and Fourier analysis which will be used throughout. In the following chapters, we shall thus assume that the reader is familiar with these notions. Nevertheless, a less advanced student will be able to find the results cited below in the book by J. Chazarain and A. Piriou [CP] (Chapter 1, Sections 1, 2 and 4) or in that by W. Rudin [R]. Readers with little knowledge of distributions are advised to read the book by L. Schwartz [S] beforehand, while students who wish to test their knowledge in this area will find a large number of exercises with solutions, together with review material in the work of C. Zuily [Z].

1. Spaces of differentiable functions and differential operators

Let Ω be an open subset of $\mathbb{R}^n$. If k is a nonnegative integer, then we let $C^k(\Omega)$ denote the space of k-times continuously differentiable functions on Ω with values in $\mathbb{C}$. Similarly, $C^\infty(\Omega)$ denotes the space of indefinitely differentiable functions on Ω. This notation extends first to the case where $\Omega = M$ is a differentiable manifold and, second, to the case where the co-domain is not $\mathbb{C}$ but a topological vector space E on $\mathbb{R}$: we then denote the corresponding spaces by $C^k(\Omega, E)$ and $C^\infty(\Omega, E)$.

For $k \in \mathbb{N} \cup \{\infty\}$, $C_0^k(\Omega)$ denotes the subspace of $C^k(\Omega)$ whose elements are zero outside a compact subset of Ω, while $C^k(\overline{\Omega})$ is formed by the restrictions to Ω of elements of $C^k(\mathbb{R}^n)$.

We shall use *multiple indices* to denote the partial derivatives of an element of $C^k(\Omega)$. A multiple index $\alpha = (\alpha_1, \dots, \alpha_n)$ is an element of $\mathbb{N}^n$, its *modulus* $|\alpha|$ is by definition $|\alpha| = \alpha_1 + \dots + \alpha_n$, and we set $\alpha! = \alpha_1! \dots \alpha_n!$. For $j \in \{1, \dots, n\}$, the derivative $\frac{\partial}{\partial x_j}$ will also be denoted by ∂_{x_j} or ∂_j when there is no risk of confusion. For reasons associated with the Fourier transformation (see Section 5 below), it is also useful to introduce the notation $D_j = -\mathrm{i}\frac{\partial}{\partial x_j}$. A higher-order derivative will then be denoted by $\partial^\alpha = \partial_1^{\alpha_1} \dots \partial_n^{\alpha_n}$ or $D^\alpha = D_1^{\alpha_1} \dots D_n^{\alpha_n}$. We shall also use this convention to denote the monomials constructed in the components of a vector of $\mathbb{R}^n$. Thus, if $x \in \mathbb{R}^n$, $x^\alpha = x_1^{\alpha_1} \dots x_n^{\alpha_n}$.

A *differential operator* on Ω is a finite linear combination of derivatives of arbitrary orders with coefficients in $C^\infty(\Omega)$. It is said to be *of order* m if it does not include derivatives of order greater than m. In other words, a differential operator of order m on Ω can be written as

$$P = \sum_{|\alpha| \le m} a_\alpha(x) D^\alpha,$$

where the $a_\alpha \in C^\infty(\Omega)$ are the coefficients of P. In this form, it is easy to see that P defines a linear mapping from $C^{k+m}(\Omega)$ to $C^k(\Omega)$ for all k. The *symbol* P is the polynomial function in ξ defined on $\Omega \times \mathbb{R}^n$ by

$$p(x, \xi) = \sum_{|\alpha| \le m} a_\alpha(x) \xi^\alpha,$$

while its *principal symbol of order* m (or *principal symbol* if there is no risk of confusion) is the homogeneous function in ξ

$$p_m(x, \xi) = \sum_{|\alpha| = m} a_\alpha(x) \xi^\alpha.$$

2. Distributions on an open set of $\mathbb{R}^n$

a) A *distribution* on an open set Ω is a linear form u on $C_0^\infty(\Omega)$ satisfying the following continuity property: for any compact subset K, there exist an integer m and a constant C such that for all $\varphi \in C_0^\infty(\Omega)$ which are zero outside K,

$$|\langle u, \varphi \rangle| \le C \sup_{x \in K} \sup_{|\alpha| \le m} |\partial^\alpha \varphi(x)|.$$

The space of distributions on Ω is denoted by $\mathcal{D}'(\Omega)$. In particular, it contains the space $L^1_{\mathrm{loc}}(\Omega)$ of *locally integrable* functions on Ω, via the following identification:

$$\forall \varphi \in C_0^\infty(\Omega),\ \forall f \in L^1_{\mathrm{loc}}(\Omega),\quad \langle f, \varphi \rangle = \int f(x) \varphi(x) \mathrm{d}x. \tag{2.1}$$

Another example of a distribution is given by the *Dirac mass* at a point. If $x_0 \in \Omega$ and $\varphi \in C_0^\infty(\Omega)$, we denote $\langle \delta_{x_0}, \varphi\rangle = \varphi(x_0)$.

b) Let $u \in \mathcal{D}'(\Omega)$. Define $\partial_j u \in \mathcal{D}'(\Omega)$ by the formula

$$\langle \partial_j u, \varphi\rangle = -\langle u, \partial_j \varphi\rangle,$$

which, taking into account the identity (2.1), extends the operator ∂_j previously defined on $C^1(\Omega)$ to distributions. Similarly, if $a \in C^\infty(\Omega)$, $au \in \mathcal{D}'(\Omega)$ is defined by

$$\langle au, \varphi\rangle = \langle u, a\varphi\rangle.$$

Thus, any differential operator $P = \sum a_\alpha D^\alpha$ extends to a linear mapping from $\mathcal{D}'(\Omega)$ to itself, via the formula

$$\langle Pu, \varphi\rangle = \langle u, {}^tP\varphi\rangle,$$

where ${}^tP\varphi = \sum(-1)^{|\alpha|} D^\alpha(a_\alpha \varphi)$.

c) If Ω' is an open subset of Ω, and if $u \in \mathcal{D}'(\Omega)$, then the *restriction* $u|_{\Omega'}$ of u to Ω' is just the restriction of the linear form u to the space $C_0^\infty(\Omega') \subset C_0^\infty(\Omega)$. Then u is said to be zero (resp. of class C^k) on Ω' if $u|_{\Omega'} = 0$ (resp. $u_{\Omega'}$ can be defined by $f \in C^k(\Omega')$, according to formula (2.1)). For this definition to be manageable, it is important that one should be able to recover u from its restrictions to the open sets of a covering of Ω. This is the object of the following lemma.

Lemma (Partitions of Unity). *Let (Ω_j) be a family of open subsets of Ω such that $\Omega = \bigcup_j \Omega_j$. Then there exists a family of functions (φ_j) such that:*

i) $\forall j, \varphi_j \in C^\infty(\Omega)$, $\operatorname{supp}(\varphi_j) \subset \Omega_j, 0 \le \varphi_j \le 1$.

ii) *For any compact subset K of Ω, $\{j, K \cap \operatorname{supp}\varphi_j \neq \emptyset\}$ is finite.*

iii) *In Ω, $\sum_j \varphi_j = 1$. (This sum is well defined, following* ii)*).*

In addition to the references already cited, the reader may refer to Exercise 6.1 of Chapter I for a proof of the above lemma (given under the hypothesis that $\overline{\Omega}_j$ is compact in Ω; the general case is an easy consequence).

Using this lemma, one can show, for example, that if $\bigcup_j \Omega_j = \Omega$ and if $u|_{\Omega_j} = 0$ (resp. $u|_{\Omega_j} \in C^k$) for all j, then $u = 0$ (resp. $u \in C^k(\Omega)$).

This leads us to the following definitions: the *support* of u (resp. *singular support* of u) is defined to be the complement in Ω of the points in the neighbourhood of which u is zero (resp. u is of class C^∞). The support of u is denoted by $\operatorname{supp} u$; the singular support of u is denoted by sing supp u. These are two closed sets, satisfying sing supp $u \subset \operatorname{supp} u$, and the preceding result can be paraphrased by the equivalences

$$\begin{aligned} u = 0 \quad &\Leftrightarrow \quad \operatorname{supp} u = \emptyset, \\ u \in C^\infty \quad &\Leftrightarrow \quad \text{sing supp } u = \emptyset. \end{aligned}$$

Finally, we note that if $u \in C^0(\Omega)$, then the support of u defined above coincides with the *closure* of $\{x \in \Omega, u(x) \neq 0\}$.

The space of distributions *with compact support* in Ω is denoted by $\mathcal{E}'(\Omega)$. It is identified with the space of linear forms on $C^\infty(\Omega)$ which are continuous for the topology defined by the semi-norms

$$\sup_{x\in K}\sup_{|\alpha|\leq m}|\partial^\alpha\varphi(x)|,$$

where K runs over the compact subsets of Ω and m runs over the integers.

3. Convolution

a) Let u and v be two C^∞ functions with compact support. We set

$$u*v(x)=\int u(y)v(x-y)\mathrm{d}y=\int u(x-y)v(y)\mathrm{d}y. \tag{3.1}$$

The function $u*v$ thus defined is C^∞ with compact support and satisfies

$$\operatorname{supp}(u*v)\subset\operatorname{supp}u+\operatorname{supp}v. \tag{3.2}$$

This is called the *convolution* of the two functions u and v.

Of course one can define the convolution of less regular functions. The most natural extension relates to summable functions: if u and v belong to $L^1(\mathbb{R}^n)$, then $u*v$ defined by (3.1) belongs to $L^1(\mathbb{R}^n)$ and we have

$$\int|u*v(x)|\mathrm{d}x\leq\int|u(x)|\mathrm{d}x\cdot\int|v(x)|\mathrm{d}x.$$

However, it is not this extension which we shall use the most frequently, but rather that described in sections b) and d) below, which relate to the cases where $u\in\mathcal{D}'(\mathbb{R}^n)$, $v\in C_0^\infty(\mathbb{R}^n)$, and then where $u\in\mathcal{D}'(\mathbb{R}^n)$, $v\in\mathcal{E}'(\mathbb{R}^n)$.

b) Let $u\in\mathcal{D}'(\mathbb{R}^n)$ and $v\in C_0^\infty(\mathbb{R}^n)$; then the formula

$$u*v(x)=\langle u,v_x\rangle,\ \text{with}\ v_x(y)=v(x-y),$$

defines a function $u*v$ of class C^∞ on $\mathbb{R}^n$. This function also satisfies

$$\partial^\alpha(u*v)=\partial^\alpha u*v=u*\partial^\alpha v, \tag{3.3}$$

$$\operatorname{supp}(u*v)\subset\operatorname{supp}u+\operatorname{supp}v. \tag{3.4}$$

c) The convolution is the basis for a very useful *regularization* procedure which we now describe.

Let $\varphi\in C_0^\infty(\mathbb{R}^n)$ be nonnegative with integral equal to 1, and let $\varepsilon>0$; we set $\varphi_\varepsilon(x)=\varepsilon^{-n}\varphi(x/\varepsilon)$. Then if $u\in\mathcal{D}'(\mathbb{R}^n)$, the family of C^∞ functions

$u_\varepsilon = u * \varphi_\varepsilon$ converges to u as ε tends to 0, in the sense that

$$\forall \psi \in C_0^\infty, \int u_\varepsilon(x)\psi(x)\mathrm{d}x \to \langle u, \psi\rangle. \tag{3.5}$$

This procedure of approximation by regular functions is of interest because the mode of convergence of u_ε to u is essentially described by the regularity of u. Thus, if $u \in C^k(\mathbb{R}^n)$, u_ε converges to u in the sense of the semi-norms $\sup_{x\in K}\sup_{|\alpha|\le k} |\partial^\alpha v(x)|$, where K runs over the compact subsets of $\mathbb{R}^n$; if $u \in L^p(\mathbb{R}^n)$ $(1 \le p < +\infty)$, the space of functions summable to the pth power, then u_ε tends to u in L^p.

Moreover, equation (3.4) shows that the support of u_ε is arbitrarily close to that of u when ε tends to 0. Using a 'cut-off function', we thus extend the regularization procedure to distributions defined on an open set Ω of $\mathbb{R}^n$, showing, for example, that $C_0^\infty(\Omega)$ is dense in $L^p(\Omega)$ if $p \in [1, +\infty[$ and in $\mathcal{D}'(\Omega)$ for the 'weak topology', i.e. in the sense of (3.5).

d) To define the convolution of two distributions, we first observe that if $u \in \mathcal{D}'(\mathbb{R}^n)$, $v, \varphi, \in C_0^\infty(\mathbb{R}^n)$, then

$$\int u * v(x)\varphi(x)\mathrm{d}x = \langle u, \tilde{v} * \varphi\rangle,$$

where we have set $\tilde{v}(x) = v(-x)$.

After having extended the operator $v \mapsto \tilde{v}$ to distributions by

$$\langle \tilde{v}, \varphi\rangle = \langle v, \tilde{\varphi}\rangle,$$

we set, for $u \in \mathcal{D}'(\mathbb{R}^n)$, $v \in \mathcal{E}'(\mathbb{R}^n)$ and $\varphi \in C_0^\infty(\mathbb{R}^n)$,

$$\langle u * v, \varphi\rangle = \langle u, \tilde{v} * \varphi\rangle.$$

We thus define a distribution $u * v$ on $\mathbb{R}^n$, which again satisfies (3.3) and (3.5), to which can be added

$$\text{sing supp } (u * v) \subset \text{sing supp } u + \text{sing supp } v. \tag{3.6}$$

For example, if $\delta = \delta_0$ denotes the Dirac mass at the origin, then for any distribution u on $\mathbb{R}^n$, we have $u * \delta = u$. The convolution of distributions is fundamental in the study of differential operators with constant coefficients. An illustration of this can be found in Section I.1.1, where the proof of the relation (3.6) is also outlined.

4. Kernels

Let Ω_1 and Ω_2 be two open subsets of $\mathbb{R}^n$, and let $K \in \mathcal{D}'(\Omega_1 \times \Omega_2)$. The equation

$$\langle A_K v, u\rangle = \langle K, u \otimes v\rangle, \tag{4.1}$$

where $u \in C_0^\infty(\Omega_1)$, $v \in C_0^\infty(\Omega_2)$, $u \otimes v(x_1, x_2) = u(x_1)v(x_2)$, defines a linear mapping $A_K : C_0^\infty(\Omega_2) \to \mathcal{D}'(\Omega_1)$, which is continuous in the following sense: for all u in $C_0^\infty(\Omega_1)$, for all compact subsets K of Ω_2, there exist a constant C and an integer m such that, $\forall v \in C_0^\infty(\Omega_2)$ with support in K,

$$|\langle A_K v, u\rangle| \leq C \sup_{x \in K} \sup_{|\alpha| \leq m} |\partial^\alpha v(x)|. \tag{4.2}$$

When $K \in L^1_{\text{loc}}(\Omega_1 \times \Omega_2)$, equation (4.1) can be written in the more familiar form

$$A_K v(x_1) = \int K(x_1, x_2) v(x_2) \mathrm{d}x_2.$$

In general, the distribution K is entirely determined by equation (4.1) and is called the *kernel* of the operator A_K.

A theorem due to L. Schwartz ensures that any operator $A : C_0^\infty(\Omega_2) \to \mathcal{D}'(\Omega_1)$, which is continuous in the sense of (4.2), has a kernel. However, we shall never use this theorem in what follows, since the operators which we shall manipulate have readily identifiable kernels.

For example, the kernel of the differential operator $P = \sum a_\alpha(x) D^\alpha$ is the distribution $K(x_1, x_2) = \sum a_\alpha(x_1) D^\alpha \delta(x_1 - x_2)$, where δ is the Dirac mass; the kernel of the operator of convolution by the distribution $u \in \mathcal{D}'(\mathbb{R}^n)$ is $K(x_1, x_2) = u(x_1 - x_2)$.

The notion of kernel thus gives rise to a more algebraic study of operators. For example, given an operator A with kernel K, it is easy to define the *transpose* operator for A (denoted by ${}^tA$), characterized by

$$\forall u \in C_0^\infty,\ \forall v \in C_0^\infty, \quad \langle Av, u\rangle = \langle {}^tAu, v\rangle.$$

It suffices to take ${}^tA$ to be the operator with kernel

$${}^tK(x_1, x_2) = K(x_2, x_1).$$

The kernel also enables us to control the supports (and the singular supports). For example,

$$\operatorname{supp}(A_K v) \subset \{x_1, \exists x_2 \in \operatorname{supp} v, (x_1, x_2) \in \operatorname{supp} K\}.$$

A similar relation exists for the singular support.

5. Fourier analysis on $\mathbb{R}^n$

a) We first introduce the space $\mathcal{S}$ of C^∞ functions which decrease rapidly on $\mathbb{R}^n$. This is the space of C^∞ functions u on $\mathbb{R}^n$, satisfying

$$\forall \alpha \in \mathbb{N}^n,\ \forall \beta \in \mathbb{N}^n, \quad \sup_{x \in \mathbb{R}^n} |x^\alpha \partial^\beta u(x)| < +\infty.$$

The space $\mathcal{S}$ is then provided with the semi-norms defined in this way, as α and β run over $\mathbb{N}^n$. If $x = (x_1, \ldots, x_n)$ is a vector of $\mathbb{R}^n$, we set

$$x| = \left(\sum_{j=1}^{n} x_j^2\right)^{1/2}.$$

Example. The function defined by $u(x) = e^{-|x|^2/2}$ belongs to $\mathcal{S}$.

b) Let $u \in \mathcal{S}$. For $\xi \in \mathbb{R}^n$, we set

$$\hat{u}(\xi) = \int e^{-ix\xi} u(x) dx, \tag{5.1}$$

where we have denoted $x\xi = \sum_{j=1}^{n} x_j \xi_j$.

We thus define a continuous linear mapping

$$\begin{aligned} \mathcal{F} : \mathcal{S} &\to \mathcal{S}, \\ u &\mapsto \hat{u}, \end{aligned}$$

called the *Fourier transformation*. The function $\mathcal{F}u = \hat{u}$ is the *Fourier transform* of u.

The operator $\mathcal{F}$ has the following fundamental properties (where we see the interest in introducing D_j):

$$\widehat{D_j u}(\xi) = \xi_j \hat{u}(\xi), \tag{5.2}$$

$$\widehat{\tau_y u}(\xi) = e^{iy\xi} \hat{u}(\xi), \tag{5.2'}$$

where $\tau_y u(x) = u(x + y)$,

$$\widehat{x_j u}(\xi) = -D_j \hat{u}(\xi), \tag{5.3}$$

$$(\widehat{e^{-ix\eta} u})(\xi) = \tau_\eta \hat{u}(\xi). \tag{5.3'}$$

c) A linear form on $\mathcal{S}$ which is continuous for the semi-norms defined in a) is said to be a *tempered* distribution on $\mathbb{R}^n$. The space of tempered distributions is denoted by $\mathcal{S}'$. In particular, by making $\mathcal{S}$ operate on itself by the formula

$$\forall u \in \mathcal{S},\ \forall v \in \mathcal{S}, \quad \langle u, v \rangle = \int u(x) v(x) dx,$$

we have $\mathcal{E}'(\mathbb{R}^n) \subset \mathcal{S}'$ and $\mathcal{S} \subset \mathcal{S}'$. Moreover, since $C_0^\infty(\mathbb{R}^n) \subset \mathcal{S}$, it is easy to prove (by regularization) that $\mathcal{S}$ is dense in $\mathcal{S}'$ (with the topology of simple convergence on $\mathcal{S}$).

We now observe that Fubini's theorem implies

(5.4) $$\forall u \in \mathcal{S},\ \forall v \in \mathcal{S}, \qquad \int \hat{u}(\xi)v(\xi)\mathrm{d}\xi = \int u(x)\hat{v}(x)\mathrm{d}x.$$

We deduce that the formula

(5.5) $$\langle \hat{u}, \varphi \rangle = \langle u, \hat{\varphi} \rangle \text{ for } u \in \mathcal{S}', \varphi \in \mathcal{S},$$

defines a linear mapping $\mathcal{F} : \mathcal{S}' \to \mathcal{S}'$, which is the unique continuous extension of $\mathcal{F} : \mathcal{S} \to \mathcal{S}'$. In particular, $\mathcal{F}$ again satisfies the relations (5.2), (5.2′) and (5.3), (5.3′).

We note that the restriction of the mapping $\mathcal{F}$ to $L^1(\mathbb{R}^n)$ is again given by formula (5.1) where, if $u \in L^1(\mathbb{R}^n)$, then $\hat{u}$ is a continuous function tending to 0 at infinity. Another example is given by the formula

$$\hat{\delta} = 1,$$

which follows directly from the definitions.

Using (5.2), we see that $\hat{1} = c\delta$ for some $c \in \mathbb{C}$. We deduce that, for all $u \in \mathcal{S}$,

$$\hat{\hat{u}}(0) = \langle \delta, \hat{\hat{u}} \rangle = \langle 1, \hat{u} \rangle = \langle \hat{1}, u \rangle = cu(0),$$

and examination of a particular case (for example $u(x) = e^{-|x|^2/2}$) leads to $c = (2\pi)^n$. Using the translation operator τ_y and the formulae (5.2′) and (5.3′), we conclude that

$$\hat{\hat{u}}(x) = (2\pi)^n u(-x).$$

In other words, if $u \in \mathcal{S}$, formula (5.1) is equivalent to

(5.6) $$u(x) = \frac{1}{(2\pi)^n} \int e^{\mathrm{i}x\xi} \hat{u}(\xi)\mathrm{d}\xi.$$

This is the *Fourier inversion formula.*

$\mathcal{F}$ is therefore an isomorphism from $\mathcal{S}$ to $\mathcal{S}$, and also from $\mathcal{S}'$ to $\mathcal{S}'$, using (5.5). Its inverse is $(2\pi)^{-n}\tilde{\mathcal{F}}$, where $\tilde{\mathcal{F}}$ is the composition of $\mathcal{F}$ and the symmetry $v \mapsto \tilde{v}$ defined on $\mathcal{S}'$ by

$$\langle \tilde{v}, \varphi \rangle = \langle v, \tilde{\varphi} \rangle \quad \text{where} \quad \tilde{\varphi}(x) = \varphi(-x).$$

d) The Fourier transform also plays a particularly important role in relation to the space $L^2(\mathbb{R}^n)$. For $f, g \in L^2(\mathbb{R}^n)$, we denote the scalar product by

$$(f, g) = \int f(x)\overline{g(x)}\mathrm{d}x.$$

Following (5.4) and the Fourier inversion formula, we have, for $f, g \in \mathcal{S}$,

(5.7) $$(\hat{f}, \hat{g}) = (2\pi)^n (f, g).$$

We deduce that $\mathcal{F}$ extends to an *automorphism* of $L^2(\mathbb{R}^n)$, such that $(2\pi)^{-n/2}\mathcal{F}$ is *unitary*. This is *Plancherel's theorem.*

e) Formulae (5.2′) and (5.3′) allow us to study the link between convolution and Fourier transformation. To describe this link, it is convenient to introduce the notion of slowly increasing C^∞ functions. A function $a \in C^\infty(\mathbb{R}^n)$ is said to be slowly increasing if

$$\forall \alpha \in \mathbb{N}^n, \exists M_\alpha \in \mathbb{N}, \exists C_\alpha > 0, \forall x \in \mathbb{R}^n, |\partial^\alpha a(x)| \leq C_\alpha(1+|x|)^{M_\alpha}.$$

For example, the polynomial functions are slowly increasing C^∞ functions and, more generally, if $u \in \mathcal{E}'(\mathbb{R}^n)$, then $\hat{u}$ is a slowly increasing C^∞ function.

If a is slowly increasing C^∞, the product by a defines a continuous linear operator on $\mathcal{S}$, and thus, by duality, extends to a continuous linear operator on $\mathcal{S}'$. In particular $a \in \mathcal{S}'$ (we identify it with $a \cdot 1$).

Finally, if $v \in \mathcal{S}'$ and $\psi \in \mathcal{S}$, the convolution $v * u$ defined by

$$v * \psi(x) = \langle v, \psi_x \rangle,$$

where $\psi_x(y) = \psi(x-y)$ is slowly increasing, C^∞. We then have the following formulae:

$$\begin{aligned} &\text{if } u \in \mathcal{S}', v \in \mathcal{E}', && \mathcal{F}(u * v) = \mathcal{F}u\mathcal{F}v, \\ &\text{if } u \in \mathcal{S}', \varphi \in \mathcal{S}', && \mathcal{F}(\varphi u) = (2\pi)^{-n}\mathcal{F}u * \mathcal{F}\varphi. \end{aligned}$$

Other situations exist where the above formulae are true (notably involving the space L^1), but we shall not use them in this book.

Chapter I

Pseudo-differential operators

In this chapter we present a self-contained description of the main elements of a theory of pseudo-differential operators: symbols, operator symbolic calculus, action in Sobolev spaces, invariance under changes of coordinates.

1. Introduction

1.1. Use of the Fourier transformation. For a slowly increasing C^∞ function $a(\xi)$ (see Section 0.5 point e)) we let $a(D)$ denote the operator defined on $\mathcal{S}'(\mathbb{R}^n)$ by

$$(a(D)u)(\xi) = a(\xi)\hat{u}(\xi).$$

The function $a(\xi)$ is called the *symbol* for the operator $a(D)$. For $u \in \mathcal{S}$, the formula

$$(a(D)u)(x) = (2\pi)^{-n} \int e^{ix\xi} a(\xi)\hat{u}(\xi) d\xi, \tag{1.1.1}$$

when compared with the Fourier inversion formula

$$u(x) = (2\pi)^{-n} \int e^{ix\xi} \hat{u}(\xi) d\xi,$$

shows that $a(D)$ only modifies the fragment $\hat{u}(\xi)e^{ix\xi}$ of 'frequency' ξ of u by multiplying its 'amplitude' $\hat{u}(\xi)$ by $a(\xi)$.

Moreover, we have the formula

$$a(D)b(D) = (ab)(D),$$

which says that the symbol for the composition of two operators is just the product of the symbols: this type of property will be referred to as 'symbolic calculus' and we shall return to it in Section 1.3. Later (with additional restrictions on the growth of $a(\xi)$), we shall call $a(D)$ a 'pseudo-differential operator with constant coefficients', thereby underlining the fact that its symbol does not depend on x.

The theory which will be developed in this chapter will enable us to associate operators $a(x, D)$ with general symbols $a(x, \xi)$ in such a way that the 'amplitude modulating' nature of the action of $a(x, D)$ emphasized above is preserved and, most importantly, that we obtain a symbolic calculus of the type

$$a(x, D)b(x, D) = (a\#b)(x, D),$$

where $a\#b$ is a certain symbol which we know how to calculate (and which is just ab if b has constant coefficients). However, the rudimentary theory of operators with constant coefficients already allows us to obtain interesting results. Consider, for example, the operator

$$P = \Delta = \partial_1^2 + \cdots + \partial_n^2 \quad \text{(the Laplacian).}$$

This is an operator of the preceding type with symbol

$$a(\xi) = -|\xi|^2.$$

We can find a distribution E satisfying $PE = \delta + \omega$ (δ is the Dirac mass at 0, $\omega \in \mathcal{S}$); it suffices to take

$$\hat{E}(\xi) = -\frac{1 - \chi(\xi)}{|\xi|^2}$$

($\chi \in C_0^\infty(\mathbb{R}^n)$, $\chi = 1$ near $\xi = 0$), since

$$\widehat{PE}(\xi) = -|\xi|^2 \hat{E}(\xi) = 1 - \chi(\xi).$$

Such a distribution E is C^∞ away from 0; indeed $\widehat{\mathrm{i}x_j E} = \partial_{\xi_j} \hat{E}$ is homogeneous of degree -3 for $|\xi|$ large, $\widehat{(-\mathrm{i}x_j)^k E} = \partial_{\xi_j}^k \hat{E}$ is of degree $-2-k$, and similarly for the derivatives of E (see also Exercise 1.1). On the other hand, if $f \in \mathcal{E}'$, we have

$$P(E * f) = f + \omega * f.$$

The distribution $u = E * f$ is thus an approximate solution of the equation

$$Pv = f,$$

with an error $\omega * f$ which is of class C^∞. Finally, we note that sing supp $u =$ sing supp f; indeed, since $Pu = f + C^\infty$, f is C^∞ where u is; on the other hand, if f is C^∞ near x_0, we can write

$$u = E * f = E * (\chi f) + E * (1 - \chi) f$$

($\chi \in C_0^\infty$ equals 1 near x_0 and is such that $\chi f \in C^\infty$): $E * \chi f \in C^\infty$, and

$$(E * (1-\chi)f)(x) = \int E(x-y)(1-\chi(y))f(y)\mathrm{d}y$$

only involves $x - y$ away from 0 (where E is C^∞) if x is close to x_0, whence $u(x) \in C^\infty$ for x near x_0.

The use of E also allows us to prove that *any* solution v of $Pv = f$ is such that

$$\text{sing supp } v = \text{sing supp } f.$$

Indeed, if $f \in C^\infty$ near x_0, suppose $\chi \in C_0^\infty$ has value 1 near x_0; then $P(\chi v)$ coincides with f near x_0 and thus is of class C^∞ near x_0. Since χv and $P(\chi v)$ are in $\mathcal{E}'$, we have

$$E * P(\chi v) = \chi v + \omega * \chi v \equiv \chi v \mod C^\infty$$

and the preceding result allows us to conclude that $\chi v \in C^\infty$ near x_0.

We have thus used the operator $\hat{E}(D)$, which is also the convolution by E, first as a right-inverse for P modulo C^∞ (to obtain an 'approximate solution' of the equation $Pu = f$), then as a left-inverse for P modulo C^∞ (to study the regularity of solutions of $Pv = f$). Such an operator is called a *parametrix* of P.

However, this way of working entirely 'from a Fourier point of view', has various disadvantages: $\operatorname{supp} u$ and sing supp u cannot be controlled precisely and the technique is not convenient for dealing with the case of operators 'with variable coefficients'.

1.2. Operators with variable coefficients. The approach of Section 1.1 for a differential operator $P = \sum a_\alpha(x)D_x^\alpha$ ($a_\alpha \in \mathcal{S}$) would lead to the formula

$$\widehat{Pu}(\xi) = (2\pi)^{-n}\sum \hat{a}_\alpha * (\xi^\alpha \hat{u}(\xi))$$

which is not very pleasant to use.

It is more clever to use the formula

$$\begin{aligned} Pu(x) &= (2\pi)^{-n}\int e^{\mathrm{i}x\xi}p(x,\xi)\hat{u}(\xi)\mathrm{d}\xi, \\ p(x,\xi) &= \sum a_\alpha(x)\xi^\alpha, \end{aligned} \tag{1.2.1}$$

since then we are under no obligation to take a polynomial function in ξ (i.e. the symbol of a differential operator) for $p(x,\xi)$: any 'reasonable' function $a = a(x,\xi)$ will do the job, and the corresponding operator will be said to be pseudo-differential with symbol $a(x,\xi)$ and $a(x,D)$ will be denoted by the capital letter A when there is no danger of confusion.

By 'reasonable' we mean that

i) $a(x,\xi)$ should have a polynomial type behaviour in ξ when $|\xi| \to +\infty$, improving with differentiation with respect to ξ; that is to say that for some m we will have

$$\forall \alpha, \quad |\partial_\xi^\alpha a(x,\xi)| \le C_\alpha (1+|\xi|)^{m-|\alpha|}.$$

ii) The variation of $a(x,\xi)$ with x must be weak enough that the distinction between the amplitude $a(x,\xi)$ and the 'phase' $e^{\mathrm{i}x\xi}$ in the integral (1.2.1) is clearly preserved. To be precise, we shall require

$$\forall \alpha, \quad |\partial_x^\alpha a(x,\xi)| \le C_\alpha (1+|\xi|)^{m}.$$

A C^∞ function satisfies properties of type i) and ii) and will be called a symbol of order m (see Section 2.1 for a precise definition). It goes without saying that the requirements i) and ii) are liable to broad modifications which will be discussed in the exercises at the end of this chapter (Exercises 2.8, 2.9, 4.12).

The effectiveness of this approach has emerged historically in relation to the problem of writing down a parametrix (see Section 1.1) for an elliptic differential operator $P(x,D)$ with variable coefficients (here, elliptic means that $p_m(x,\xi) \ne 0$ for $\xi \in \mathbb{R}^n$, $\xi \ne 0$). Indeed, if one tries a solution of the form $u = a(x,D)f$, for the equation $Pu = f$, one finds that

$$Pu(x) = (2\pi)^{-n} \int e^{\mathrm{i}x\xi} p(x,\xi+D) a(x,\xi) \hat{f}(\xi) \mathrm{d}\xi,$$

with

$$p(x,\xi+D)a = p(x,\xi)a(x,\xi) + \sum_{\alpha\ne 0} \frac{1}{\alpha!} \partial_\xi^\alpha p(x,\xi) D^\alpha a.$$

In the above sum, the behaviour (when $|\xi| \to +\infty$) of the successive terms improves as one goes on: if one chooses $a(x,\xi) = (1-\chi(\xi))/p(x,\xi)$ (since $p(x,\xi) \ne 0$ for $|\xi|$ sufficiently large), one can check that a is a symbol of order $-m$, and one obtains $p(x,\xi+D)a = 1 + O(1/|\xi|) = 1 + r$, where r is a symbol of order -1. Thus $Pu = f + Rf$, where Rf is a more regular function than f (for example, if $f \in L^2$, Rf and its gradient are in L^2). As we shall see (Section 5.4), this procedure can be refined to end up with $Rf \in C^\infty$, as in Section 1.1.

The important thing here is to construct a framework (that of symbols and the corresponding operators) in which it is possible to invert (approximately) an operator $P(x,D)$ (whether differential or pseudo-differential, it matters little) by an operator $a(x,D)$ of the same type: one will then have an operator algebra; moreover, the inverse will essentially have for its symbol the 'natural candidate' $a(x,\xi) = 1/p(x,\xi)$.

We note that the approximation considered here does not arise in the sense of the numerical values of the functions, but in the sense of their regularity: a 'small error' is a C^∞ function, however large it may be! One might fear that this characteristic might limit the possible applications of the theory to the study of the singularities of the functions, in the spirit of the property proved in Section 1.1:

$$\text{if } \Delta v = f, \text{ then sing supp } v = \text{sing supp } f.$$

In fact, this is far from the case: in Chapter II, we shall see how to use pseudo-differential operators to establish inequalities between solutions and known conditions in certain differential problems; the approximate nature of the theory is reflected in the difficulty of putting precise numerical values on the constants in the inequalities; nevertheless, the very existence of such inequalities in general suffices to prove the existence of *exact* solutions to the problems considered (see Chapter II, Section C.1.2).

1.3. The two sides reconciled. The remarkable aspect of the theory outlined here is the great robustness of the notion of the symbol. For example, the product $a(x,D)b(x,D)$ of two operators associated by (1.2.1) with symbols $a(x,\xi)$ and $b(x,\xi)$ is indeed formally an operator of the same form associated with some function $c(x,\xi)$; the fact that $c(x,\xi)$ is itself a symbol is highly non-trivial and constitutes the very core of the theory of this chapter (Theorem 4.1). We also have a 'symbolic calculus' in the sense of Section 1.1, in that the symbol c can be written asymptotically (notion defined in Section 2)

$$c \sim \sum_{\alpha} \frac{1}{\alpha!} \partial_\xi^\alpha a D_x^a b,$$

where the principal term of this sum is ab, as in the case of constant coefficients.

Such a symbolic calculus is of interest in that it reduces statements relating to operators to statements relating to functions: we shall see numerous examples in this chapter (notably in Section 5).

Moreover, one consequence of the theory is the following 'pseudo-local' property: for any pseudo-differential operator A,

$$\text{sing supp } Au \subset \text{sing supp } u \quad \text{(compare with Section 1.1).}$$

This means that the action of an operator A, while it 'spreads out' the support of u, does not however ruin the localization in x of the *singularities* of u (which fact is nowhere evident in formula (1.2.1)).

Refining this, we arrive (see Section II.B) at the concept of the *wave front set* of u (denoted by $WF(u)$), which is a set of points (x,ξ) where $x \in$ sing supp u while ξ is a *dominant frequency* of u near x and at the

formula

$$WF(Au) \subset WF(u).$$

We thus see that the pseudo-differential calculus and the associated concept of the wave front set provide double the information about the functions being considered, 'from the x viewpoint' and from the 'Fourier viewpoint'. The systematic usage of corresponding localizations 'in x' and 'in ξ', known as microlocal analysis, will be outlined in Sections II.A and II.B.

2. Symbols

2.1. Definition and examples.

Definition. *Let $m \in \mathbb{R}$. Let $S^m = S^m(\mathbb{R}^n \times \mathbb{R}^n)$ denote the set of $a \in C^\infty(\mathbb{R}^n \times \mathbb{R}^n)$ such that*

$$\forall\alpha, \forall\beta,\ |\partial_x^\alpha \partial_\xi^\beta a(x,\xi)| \leq C_{\alpha,\beta}(1+|\xi|)^{m-|\beta|}. \tag{2.1.1}$$

We also denote $S^{-\infty} = \bigcap_m S^m$. An element $a \in S^m$ is called a symbol of order m.

Example 1. If $a(x,\xi) = \sum_{|\alpha|\leq m} a_\alpha(x)\xi^\alpha$, with $a_\alpha \in C^\infty(\mathbb{R}^n)$, is bounded together with all its derivatives, and $a \in S^m$, then a is said to be a differential symbol.

Example 2. Let $a(\xi)$ be a (positively) homogeneous function of degree m (that is, $\forall\lambda > 0$, $a(\lambda\xi) = \lambda^m a(\xi)$) which is C^∞ for $\xi \neq 0$. If $\chi \in C_0^\infty$, with $\chi = 1$ near 0, then the function $\tilde{a}(\xi) = (1-\chi(\xi))a(\xi)$ is a symbol of order m.

Example 3. Let $p(x,\xi)$ be a differential symbol of order m defined for $x \in \Omega$ where Ω is an open subset of $\mathbb{R}^n$. Suppose that p is elliptic in Ω, that is,

$$\forall x \in \Omega, \forall\xi \in \mathbb{R}^n, \xi \neq 0,\ p_m(x,\xi) \neq 0.$$

For $\varphi \in C_0^\infty(\Omega)$ and $\chi \in C_0^\infty$, with $\chi = 1$ near 0, the function $a(x,\xi) = \varphi(x)(1-\chi(\xi/C))/p(x,\xi)$ is a symbol of order $-m$ if C is sufficiently large (see the discussion in Section 1.3).

Example 4. Let $\varphi \in \mathcal{S}$. Then the function $\varphi(\xi)$ is a symbol of order $-\infty$.

Example 5. The function $a(x,\xi) = e^{\mathrm{i}x\xi}$ is not a symbol.

We note the following elementary properties:

$$\text{if } a \in S^m, \text{ then } \partial_x^\alpha \partial_\xi^\beta a \in S^{m-|\beta|}, \tag{2.1.2}$$

$$\text{if } a \in S^m \text{ and } b \in S^{m'}, \text{ then } ab \in S^{m+m'}, \tag{2.1.3}$$

$$\text{if } a \in S^m, \text{ then } a \in \mathcal{S}'(\mathbb{R}^{2n}). \tag{2.1.4}$$

Finally, the following lemma is useful.

Lemma 2.1.1. *If $a_1, \dots, a_k \in S^0$ and $F \in C^\infty(\mathbb{C}^k)$, then $F(a_1, \dots, a_k) \in S^0$.*

Proof. Since $\operatorname{Re} a_k$ and $\operatorname{Im} a_k$ are in S^0, we may suppose that the a_i are real valued and $F \in C^\infty(\mathbb{R}^k)$. Now

$$\frac{\partial}{\partial x_j} F(a) = \sum \frac{\partial F}{\partial a_k} \partial_{x_j} a_k, \tag{2.1.5}$$

$$\frac{\partial}{\partial \xi_j} F(a) = \sum \frac{\partial F}{\partial a_k} \partial_{\xi_j} a_k. \tag{2.1.6}$$

We shall now prove by induction on p that the estimates of type (2.1.1) for any function of type $F(a)$ are valid for $|\alpha| + |\beta| \le p$. The case $p = 0$ is clear. Now, for $|\alpha| + |\beta| \le p + 1$, Leibniz's formula applied to (2.1.5) or (2.1.6) and the induction hypothesis applied to the derivatives of $\frac{\partial F}{\partial a_k}(a)$ give the desired result. □

One should be aware that a symbol $a(x, \xi)$ ($x \in \mathbb{R}^n$, $\xi \in \mathbb{R}^n$) is not a symbol in the variables $y = (x, x')$, $\eta = (\xi, \xi')$ ($x' \in \mathbb{R}^m$, $\xi' \in \mathbb{R}^m$, $m \ge 1$), except when it is differential (Exercise 2.1). Numerous other classes of symbols are also useful in practice: some of these are introduced in Exercises 2.8 and 2.9.

2.2. Approximation of symbols. Let us define the semi-norms on S^m:

$$|a|^m_{\alpha,\beta} = \sup_{(x,\xi) \in \mathbb{R}^n \times \mathbb{R}^n} \left\{ (1 + |\xi|)^{-(m - |\beta|)} |\partial_x^\alpha \partial_\xi^\beta a(x, \xi)| \right\}.$$

We thus obtain a complete space (to be precise, this is called a Fréchet space), where the convergence $a_n \to a$ means that $\forall \alpha, \forall \beta, |a_n - a|^m_{\alpha,\beta} \to 0$. It is clear that for this topology, the mappings defined in (2.1.2) and (2.1.3) are continuous.

It is convenient to have the following approximation lemma.

Lemma 2.2.1. *Let $a \in S^0(\mathbb{R}^n \times \mathbb{R}^n)$, and set $a_\varepsilon(x, \xi) = a(x, \varepsilon\xi)$. Then a_ε is bounded in S^0, and $a_\varepsilon \to a_0$ in S^m for all $m > 0$, when $\varepsilon \to 0$.*

Proof. We shall show that, for $0 \le m \le 1$, $0 \le \varepsilon \le 1$, α and β arbitrary, $|a_\varepsilon - a_0|^m_{\alpha,\beta} \le C_{\alpha,\beta,m} \varepsilon^m$. Indeed, for $\beta = 0$,

$$\partial_x^\alpha (a_\varepsilon - a_0) = \varepsilon\xi \cdot \int_0^1 \partial_\xi \partial_x^\alpha a(x, t\varepsilon\xi) \mathrm{d}t,$$

whence

$$|\partial_x^\alpha (a_\varepsilon - a_0)| \le C \int_0^{\varepsilon|\xi|} \frac{\mathrm{d}s}{1+s} = C \log(1 + \varepsilon|\xi|),$$

and the estimate follows from $\log(1 + x) \le C_m x^m$ ($x \ge 0$, $m > 0$).

For $\beta \neq 0$, $\partial_x^\alpha \partial_\xi^\beta a_0 = 0$, and

$$|\partial_x^\alpha \partial_\xi^\beta a_\varepsilon| \leq C_{\alpha,\beta} \varepsilon^{|\beta|} (1 + \varepsilon|\xi|)^{|\beta|},$$

which gives the result, since $(1 + |\xi|) \geq 1 + \varepsilon|\xi|$.

We note that the convergence $a_\varepsilon \to a_0$ does not take place in S^0. □

In particular, if $\chi \in \mathcal{S}$, $\chi = 1$ near 0, and $a \in S^m$, then the symbols $a_\varepsilon(x,\xi) = \chi(\varepsilon\xi)a(x,\xi)$ are of order $-\infty$, and $a_\varepsilon \to a$ in $S^{m'}$ for all $m' > m$.

2.3. Asymptotic sums and classical pseudo-differential symbols in $\mathcal{S}$ and $\mathcal{S}'$. Let $a_j \in S^{m_j}$ ($j \in \mathbb{N}$) for a decreasing sequence $m_j \to -\infty$ (in practice, this will often be $m_j = m - j$ or $m_j = m - j/2$). We wish to give a meaning to $\sum_j a_j$, it being understood that we cannot hope for a convergent series here; we shall write

$$a \sim \sum a_j$$

(asymptotic sum in the sense of the behaviour when $|\xi| \to +\infty$) to mean that

$$\forall k \geq 0, \quad a - \sum_{j=0}^{k} a_j \in S^{m_{k+1}}.$$

To illustrate the notion of the asymptotic sum, let us first prove the following classical lemma.

Lemma (Borel). *Let (b_j) be a sequence of complex numbers. There exists $f \in C^\infty(\mathbb{R})$ such that $\forall j$, $f^{(j)}(0) = b_j$, that is $f(x) \sim \sum b_j \frac{x_j}{j!}$ when $x \to 0$.*

Proof. Let χ be a C^∞ function equal to 1 for $|x| \leq 1$ and 0 for $|x| \geq 2$.

Let (λ_j) be a sequence of positive numbers tending to $+\infty$. We shall show that we can choose (λ_j) such that the function defined by

$$f(x) = \sum_{j=0}^{\infty} b_j \frac{x^j}{j!} \chi(\lambda_j x)$$

has the properties of the statement. Let us first note that the above series is simply convergent. Let k be an integer. If $j \geq k$, the kth derivative of the term of rank j is equal to

$$f_j^{(k)}(x) = \sum_{0 \leq l \leq k} \binom{k}{l} b_j \frac{x^{j-l}}{(j-l)!} x^{(k-l)}(\lambda_j x) \lambda_j^{k-l}.$$

Since $\lambda_j x$ is bounded on the supports of χ and its derivatives, there exists a constant C_k such that

$$\left| f_j^{(k)}(x) \right| \leq C_k |b_j| \lambda_j^{k-j} \frac{1}{(j-k)!}.$$

If we choose $\lambda_j \geq 1 + |b_j|$, then the series $\sum_j |f_j^{(k)}(x)|$ converges uniformly for $x \in \mathbb{R}$, which ensures that f is of class C^∞ and that its derivatives are obtained by differentiating term by term. In particular, for all k,

$$f^{(k)}(0) = b_k.$$

$\square$

The same lemma is true for $f = f(z)$ holomorphic in a sector of $\mathbb{C}$ (Exercise 2.6).

In the case of an asymptotic sum of symbols, we have the following proposition:

Proposition 2.3. *There exists $a \in S^{m_0}$ such that $a \sim \sum a_j$. Moreover we can insist that* $\operatorname{supp} a \subset \bigcup \operatorname{supp} a_j$.

Proof. This is an adaptation of the preceding lemma, with the asymptotics for $x \to 0$ replaced by the asymptotics for $1/|\xi| \to 0$. Thus we take

$$a = \sum \tilde{a}_j = \sum (1 - \chi(\varepsilon_j \xi)) a_j,$$

with $\chi \in C_0^\infty$, $\chi = 1$ near 0, and $\varepsilon_j \downarrow 0$ sufficiently rapidly. More precisely, we shall require

$$\left| \partial_x^\alpha \partial_\xi^\beta \tilde{a}_j \right| \leq 2^{-j} (1 + |\xi|)^{1 + m_j - |\beta|} \quad \text{if } |\alpha| + |\beta| \leq j,$$

which is possible by virtue of the Approximation Lemma 2.2.1 since $1 - \chi(\varepsilon\xi)$ tends to 0 in S^1. The sum is locally finite, thus $a \in C^\infty$.

For α, β, k given, we have, for $N \geq |\alpha| + |\beta|$ and $m_N + 1 \leq m_{k+1}$,

$$\left| \partial_x^\alpha \partial_\xi^\beta \left(a - \sum_{j \leq N-1} \tilde{a}_j \right) \right| \leq (1 + |\xi|)^{m_{k+1} - |\beta|};$$

hence

$$a - \sum_{j \leq k} a_j = a - \sum_{j \leq N-1} \tilde{a}_j + \sum_{k+1 \leq j \leq N-1} \tilde{a}_j + \sum_{j \leq k} (a_j - \tilde{a}_j)$$

satisfies

$$\left| \partial_x^\alpha \partial_\xi^\beta \left(a - \sum_{j \leq k} a_j \right) \right| \leq C_{\alpha,\beta,k} (1 + |\xi|)^{m_{k+1} - |\beta|},$$

since $a_j - \tilde{a}_j \in S^{-\infty}$ and $\tilde{a}_j \in S^{m_j}$. $\square$

In the same order of ideas, we introduce the following definition.

Definition. *A symbol $a \in S^m$ is said to be classical if $a \sim \sum_j a_j$, where the functions a_j are homogeneous of degree $m-j$ for $|\xi| \geq 1$, that is $a_j(x, \lambda\xi) = \lambda^{m-j} a_j(x,\xi)$ for $|\xi| \geq 1$, $\lambda \geq 1$.*

The first three examples given in Section 2.1 are classical.

Sometimes, by abuse, functions a which are C^∞ for $|\xi| \neq 0$ only are classified as 'symbols': it is understood that these have to be cut off appropriately near $\xi = 0$, where the difference between two symbols obtained in this way is then in $S^{-\infty}$. This is, in particular, the case for homogeneous functions.

3. Pseudo-differential operators in $\mathcal{S}$ and $\mathcal{S}'$

We now make the definition mentioned in the introduction precise.

3.1. Action in $\mathcal{S}$.

Proposition 3.1. *If $a \in S^m$ and $u \in \mathcal{S}$, the formula*

$$\mathrm{Op}(a)u(x) = (2\pi)^{-n} \int e^{\mathrm{i}x\xi} a(x,\xi)\hat{u}(\xi)\mathrm{d}\xi$$

defines a function of $\mathcal{S}$, and the mapping

$$(a, u) \mapsto \mathrm{Op}(a)u$$

is continuous. The linear mapping Op *defined in this way from S^m to the linear operators of $\mathcal{S}$ is injective and satisfies the relations*

$$\begin{aligned} [\mathrm{Op}(a), D_j] &= \mathrm{i}\mathrm{Op}(\partial_{x_j} a), \\ [\mathrm{Op}(a), x_j] &= -\mathrm{i}\mathrm{Op}(\partial_{x_j} a), \end{aligned} \tag{3.1.1}$$

where x_j denotes multiplication by the function $x \mapsto x_j$.

Proof. First, since $\hat{u} \in \mathcal{S}$ and $a \in S^m$ we have

$$\begin{aligned} |\mathrm{Op}(a)u(x)| \leq\ & (2\pi)^{-n}(\sup |a(x,\xi)(1+|\xi|)^{-m}|) \\ & \times \int (1+|\xi|)^m |\hat{u}(\xi)|\mathrm{d}\xi, \end{aligned}$$

which shows that $\mathrm{Op}(a)u$ is (continuous and) bounded.

The relations (3.1.1) follow immediately by integration by parts. For example,

$$\begin{aligned} \mathrm{Op}(a)D_j u(x) &= (2\pi)^{-n} \int e^{\mathrm{i}x\xi} \widehat{D_j u}(\xi)\mathrm{d}\xi \\ &= (2\pi)^{-n} \int e^{\mathrm{i}x\xi} a(x,\xi)\xi_j \hat{u}(\xi)\mathrm{d}\xi, \end{aligned}$$

whilst

$$D_j(\mathrm{Op}(a)u)(x) \quad = -\mathrm{i}\left[(2\pi)^{-n} \textstyle\int e^{\mathrm{i}x\xi}\mathrm{i}\xi_j a(x,\xi)\hat{u}(\xi)\mathrm{d}\xi + \mathrm{Op}(\partial_{x_j}a)u(x)\right],$$

whence we have the first formula of (3.1.1). These formulae imply, among other things, that $x^\alpha D^\beta(\mathrm{Op}(a)u)$ is a linear combination of terms

$$\mathrm{Op}(\partial_\xi^{\alpha'}\partial_x^{\beta'}a)(x_{\alpha''}D^{\beta''}u), \text{ with } \alpha'+\alpha''=\alpha, \beta'+\beta''=\beta;$$

hence $x^\alpha D^\beta(\mathrm{Op}(a)u)$ is bounded by the product of a semi-norm of u in $\mathcal{S}$ by a semi-norm of a in S^m, which is the continuity result stated.

Finally, to show that Op is injective, let us suppose that, for all $u \in \mathcal{S}$, and for all $x \in \mathbb{R}^n$, we have

$$\int e^{\mathrm{i}x\xi}a(x,\xi)\hat{u}(\xi)\mathrm{d}\xi = 0.$$

Then, fixing x, we deduce that the function b defined by

$$b(\xi) = \frac{a(x,\xi)}{(1+|\xi|^2)^{\frac{m}{2}+\frac{n}{4}+\frac{1}{2}}}$$

belongs to $L^2(\mathbb{R}^n)$ and is orthogonal to all the functions of the form

$$v(\xi) = e^{-\mathrm{i}x\xi}(1+|\xi|^2)^{\frac{m}{2}+\frac{n}{4}+\frac{1}{2}}\overline{\hat{u}(\xi)};$$

when u describes $\mathcal{S}$, v describes $\mathcal{S}$, thus $b = 0$ by density of $\mathcal{S}$ in L^2. □

Definition. *For $a \in S^m$, the operator* Op(a) *is the pseudo-differential operator with symbol a. A pseudo-differential operator is said to be of order m if its symbol belongs to S^m.*

Notation By analogy with differential operators (see formula (1.2.1)), we often denote $\mathrm{Op}(a) = a(x,D)$. When there can be no confusion, a pseudo-differential operator will also sometimes be denoted by the capital letter corresponding to its symbol; thus $A = \mathrm{Op}(a)$.

3.2. Kernels and adjoints.

3.2.1. *Kernels.* First, let $a \in S^{-\infty}$. For $u \in \mathcal{S}$, we have

$$\begin{aligned}
\mathrm{Op}(a)u(x) &= (2\pi)^{-n}\int e^{\mathrm{i}x\xi}a(x,\xi)\hat{u}(\xi)\mathrm{d}\xi \\
&= (2\pi)^{-n}\int e^{\mathrm{i}x\xi}a(x,\xi)\mathrm{d}\xi \int e^{-\mathrm{i}y\xi}u(y)\mathrm{d}y \\
&= (2\pi)^{-n}\int u(y)\mathrm{d}y \int e^{\mathrm{i}(x-y)}\xi a(x,\xi)\mathrm{d}\xi.
\end{aligned}$$

The kernel K of the operator Op(a) is thus given by

$$K(x,y) = (2\pi)^{-n}\int e^{\mathrm{i}(x-y)}\xi a(x,\xi)\mathrm{d}\xi. \tag{3.2.1}$$

When a belongs to S^m, this formula extends as

$$K(x,y) = (2\pi)^{-n}(\mathcal{F}_\xi a)(x, y-x) \tag{3.2.2}$$

where $\mathcal{F}_\xi a$ denotes the Fourier transform of a with respect to the variable ξ, in $\mathcal{S}'(\mathbb{R}^{2n})$. The Fourier inversion formula then gives

$$a(x,\xi) = \mathcal{F}_{y\to\xi}[K(x, x-y)]. \tag{3.2.2'}$$

The two formulae (3.2.2) and (3.2.2′) establish a bijection in $\mathcal{S}'(\mathbb{R}^{2n})$ between symbols and operator kernels. Of course if a were only in $\mathcal{S}'(\mathbb{R}^{2n})$, the corresponding operator would map $\mathcal{S}$ to $\mathcal{S}'$, and not $\mathcal{S}$ to $\mathcal{S}$ as shown in Proposition 3.1. We deduce that K has additional properties which characterize it among tempered distributions as the kernel of a pseudo-differential operator (see Exercises 3.1 and 3.2). We shall not use these in this book.

3.2.2. *Adjoints.* Let A be an arbitrary operator from $\mathcal{S}$ to $\mathcal{S}$. We seek an operator A^* from $\mathcal{S}$ to $\mathcal{S}$ such that

$$\forall u \in \mathcal{S}, \forall v \in \mathcal{S},\ (Au, v) = (u, A^*v).$$

We note that, by a simple density argument which we have already met, if A^* exists, it is unique; A^* is then called the *adjoint* of A.

The existence of A^* allows us to define $A : \mathcal{S}' \to \mathcal{S}'$ by the formula

$$\forall u \in \mathcal{S}', \forall v \in \mathcal{S},\ (Au, v) = (u, A^*v),$$

it being understood that, if $u \in \mathcal{S}'$ and $v \in \mathcal{S}$, then (u, v) denotes $\langle u, \overline{v}\rangle$, where $\langle\cdot,\cdot\rangle$ is the duality bracket: if we prefer, we will then have

$$\langle Au, v\rangle = \langle u, \overline{A^*\overline{v}}\rangle. \tag{3.2.3}$$

This type of manipulation has been used, in particular, to define the Fourier transformation on $\mathcal{S}'$, since we have

$$\mathcal{F}^* = \overline{\mathcal{F}}.$$

Let us now study these problems for two simple examples of pseudo-differential operators:

a) If $P = \sum_{|\alpha|\le m} a_\alpha(x)D^\alpha$ is a differential operator with slowly increasing C^∞ coefficients, then for all functions u, v in $\mathcal{S}$ we have

$$(Pu, v) = (u, P^*v),$$

where $P^*v = \sum_{|\alpha|\le m} D^\alpha(\overline{a}_\alpha v)$.

Indeed, $(au, v) = (u, \overline{a}v)$ and

$$\begin{aligned}(D_j u, v) &= \int D_j u\overline{v}\mathrm{d}x = -\mathrm{i}\int \partial_j u\overline{v}\mathrm{d}x \\ &= \mathrm{i}\int u\partial_j\overline{v}\mathrm{d}x = \int u\overline{D_j v}\mathrm{d}x = (u, D_j v).\end{aligned}$$

P^* is thus a differential operator with slowly increasing coefficients in C^∞. We note that its principal symbol is simply the *conjugate* of the principal symbol of P.

b) Let $a(D)$ be a pseudo-differential operator with constant coefficients (i.e. with symbol a independent of x). For u and v in $\mathcal{S}$ we have

$$(a(D)u, v) = (2\pi)^{-n}(a\hat{u}, \hat{v}) = (2\pi)^{-n}(\hat{u}, \overline{a}\hat{v}) = (u, \overline{a}(D)v).$$

Thus $a(D)^* = \overline{a}(D)$.

In the two cases a) and b), the adjoint of the pseudo-differential operator considered is a pseudo-differential operator of the same order. This is true in general, but the proof is long.

To end, let us observe that if A^* exists, there is an easy expression for its kernel K^* using the kernel K of A. Indeed, following the definition of the kernel of an operator and formula (3.2.3), we have

$$\begin{aligned}\langle K(x,y), u(y)v(x)\rangle &= \langle Au, v\rangle = \langle u, \overline{A^*\overline{v}}\rangle\\ &= \overline{\langle \overline{u}, A^*, \overline{v}\rangle}\\ &= \overline{\langle K^*(y,x), \overline{v}(x)\overline{u}(y)\rangle}.\end{aligned}$$

Finally

$$K^*(y,x) = \overline{K(x,y)}. \tag{3.2.4}$$

3.2.3. *Adjoint of a pseudo-differential operator.* Suppose we have a pseudo-differential operator A with symbol $a \in S^m$. We would like to determine the adjoint of A, if it exists. For this, in conformance with the two preceding sections, it suffices to verify that if K is the kernel of a (given by formula (3.2.2)), the operator with kernel K^* given by (3.2.4) sends $\mathcal{S}$ to $\mathcal{S}$. We shall in fact show that it is a pseudo-differential operator. For this, we need to show that the distribution a^* in $\mathcal{S}'(\mathbb{R}^{2n})$ corresponding to K^* by (3.2.2$'$) is a symbol. To establish the formula giving a^* as a function of a, it is more convenient to assume that a (and hence a^*) is in $\mathcal{S}(\mathbb{R}^{2n})$ and then to extend by continuity to $\mathcal{S}'(\mathbb{R}^{2n})$, by virtue of the continuity of the operations (3.2.2), (3.2.2$'$) and (3.2.4). We obtain

$$K^*(x,y) = \overline{K(y,x)} = (2\pi)^{-n}\int e^{i(x-y)\xi}\overline{a(y,\xi)}d\xi$$

and

$$\begin{aligned}a^*(x,\xi) &= \int K^*(x, x-y)e^{-iy\xi}dy\\ &= (2\pi)^{-n}\int e^{iy(\eta-\xi)}\overline{a}(x-y,\eta)dyd\eta\\ &= (2\pi)^{-n}\int e^{-iy\eta}\overline{a}(x-y,\xi-\eta)dyd\eta.\end{aligned}$$

The correspondence $a \mapsto a^*$ is thus a convolution with the oscillating exponential $(2\pi)^{-n}e^{-iy\eta}$. It is easy to verify directly that this operation sends $\mathcal{S}'(\mathbb{R}^{2n})$ to $\mathcal{S}'(\mathbb{R}^{2n})$, since (see Exercise 4.9 a)) the Fourier transform of $(2\pi)^{-n}\exp(-ix\xi)$ at (x,ξ) is nothing other than $\exp(i\hat{x}\hat{\xi})$ which is a slowly increasing C^∞ function, and thus a multiplier for $\mathcal{S}'(\mathbb{R}^{2n})$. The crucial point is given by the following theorem which is one of the two fundamental results of the symbolic calculus.

Theorem 3.2.3. *If $a \in S^m$, $a^* \in S^m$ and*

$$a^*(x,\xi) \sim \sum_\alpha \frac{1}{\alpha!}\partial_\xi^\alpha D_x^\alpha \overline{a}(x,\xi).$$

In particular, if $A = \mathrm{Op}(a)$ is a pseudo-differential operator of order m, $A^* = \mathrm{Op}(a^*)$ is a pseudo-differential operator of order m and, consequently, A extends to an operator from $\mathcal{S}'(\mathbb{R}^n)$ to $\mathcal{S}'(\mathbb{R}^n)$.

The proof of the theorem is based on delicate estimates; although this is central to the subject, readers with heavy demands on their time may neglect the details, which is why we present them in an appendix. (See also the alternative proof proposed in Exercise 4.11).

4. Composition of operators

Let A_1 and A_2 be two pseudo-differential operators. For $u \in \mathcal{S}$ we have

$$A_1A_2(u)(x) = (2\pi)^{-n}\int e^{ix\xi}a_1(x,\xi)\widehat{A_2u}(\xi)\mathrm{d}\xi$$

and

$$\widehat{A_2u}(\xi) = (2\pi)^{-n}\int e^{-iy(\xi-\eta)}a_2(y,\eta)\hat{u}(\eta)\mathrm{d}y\mathrm{d}\eta,$$

hence

$$A_1A_2u(x) = (2\pi)^{-n}\int e^{iy\eta+i\xi(x-y)}a_1(x,\xi)a_2(y,\eta)\hat{u}(\eta)\mathrm{d}\xi\mathrm{d}\eta\mathrm{d}y;$$

formally, the operator A_1A_2 is equal to B, where

$$b(x,\xi) = (2\pi)^{-n}\int e^{-i(x-y)(\xi-\eta)}a_1(x,\eta)a_2(y,\xi)\mathrm{d}y\mathrm{d}\eta. \tag{4.1}$$

Exactly as in Section 3.2.3, the integral defining b is a convolution in the variables (y,η) (for (x,ξ) fixed). The same proof as in Section 3.2.3 gives the second fundamental theorem of the symbolic calculus.

Theorem 4.1. *If $a_1 \in S^{m_1}$, $a_2 \in S^{m_2}$, we have $\mathrm{Op}(a_1)\mathrm{Op}(a_2) = \mathrm{Op}(b)$, where $b = a_1\#a_2 \in S^{m_1+m_2}$ is given by (4.1), and $b \sim \sum_\alpha \frac{1}{\alpha!}\partial_\xi^\alpha a_1 D_x^\alpha a_2$.*

Of course, in the case where a_1 and a_2 are differential, one can verify the theorem using Leibniz's formula, the asymptotic formula then being exact (Exercise 4.1).

Corollary 4.1. *If $a_1 \in S^{m_1}$ and $a_2 \in S^{m_2}$, the bracket $[A_1, A_2] = A_1A_2 - A_2A_1$ for the operators A_1 and A_2 is an operator of order $m_1 + m_2 - 1$ and its symbol b is equal to*

$$b = \frac{1}{i}\{a_1, a_2\} \mod S^{m_1+m_2-2}.$$

Here, we have used the notation

$$\{f, g\} = \sum \left(\frac{\partial f}{\partial \xi_j} \frac{\partial g}{\partial x_j} - \frac{\partial f}{\partial x_j} \frac{\partial g}{\partial \xi_j} \right),$$

the Poisson bracket of the two functions $f(x, \xi), g(x, \xi)$.

Proof of Corollary 4.1. The symbols b_1 and b_2 for A_1A_2 and A_2A_1 satisfy

$$\begin{aligned} b_1 &\sim \sum \frac{1}{\alpha!} \partial_\xi^\alpha a_1 D_x^\alpha a_2, \\ b_2 &\sim \sum \frac{1}{\alpha!} \partial_\xi^\alpha a_2 D_x^\alpha a_1. \end{aligned}$$

Thus, for $b = b_1 - b_2$,

$$\begin{aligned} b &= \sum_j \left(\frac{\partial a_1}{\partial \xi_j} D_j a_2 - \frac{\partial a_2}{\partial \xi_j} D_j a_1 \right) \mod S^{m_1+m_2-2} \\ &= \frac{1}{i}\{a_1, a_2\} \mod S^{m_1+m_2-2}. \end{aligned}$$

□

In Section 6 we shall see how the principal symbol of an operator A on a manifold M is defined as a function on the *cotangent bundle* T^*M of that manifold, where the Poisson bracket of two such functions is defined intrinsically (that is, without reference to a choice of local coordinates (x, ξ)).

Theorem 4.1 constitutes the 'symbolic calculus' we referred to in the introduction (an alternative proof is given in Exercises 4.10 and 4.11).

5. Action of pseudo-differential operators and Sobolev spaces

5.1. Action on L^2.

Theorem 5.1. *If $a \in S^0$, $a(x, D)$ is an endomorphism of L^2.*

We remark that this fundamental result is self-evident if the symbol $a(x,\xi)$ depends only on x or only on ξ, or again if $a(x,\xi) = b(x)c(\xi)$: in these special cases, the norm of the operator can be easily calculated as a function of $\sup|a|$.

The general case, on the other hand, is much more complicated (as is the calculation of the norm) and requires control of a certain number of derivatives of the symbol a.

Here, we give a proof using the symbolic calculus: other variants are examined in the exercises (Exercises 5.2 and 5.3).

Henceforth we shall use $|u|_0$ to denote the norm of u in L^2.

Proof. The idea is as follow. Since $|Au|_0^2 = (Au, Au) = (A^*Au, u)$, the inequality $|Au|_0^2 \leq M|u|_0^2$ can also be written as $(Bu, u) \geq 0$, where $B = M - A^*A$ is self-adjoint of order 0. To prove that B satisfies $(Bu, u) \geq 0$ for M sufficiently large, the simplest thing is again to express B in the form $B = C^*C$, which is what we do.

a) We choose $M \geq 2\sup|a(x,\xi)|^2$ and take

$$c(x,\xi) = (M - |a(x,\xi)|^2)^{1/2}.$$

Following Lemma 2.1.1, $c \in S^0$; Theorem 4.1 implies $C^*C = M - A^*A + R$, $r \in S^{-1}$ (as usual, our programme is only realized approximately). Thus $|Au|_0^2 \leq M|u|_0^2 + (Ru, u)$.

b) We must now determine an upper bound for the 'error' (Ru, u). Let us suppose that $r \in S^{-k}$ with $k \geq 1$: since $|Ru|_0^2 = (R^*Ru, u)$, R will be continuous in L^2 if R^*R is, with $\|R\|_{L^2\to L^2} \leq \|R^*R\|_{L^2\to L^2}^{1/2}$. Now (trick!) $r^*\#r \in S^{-2k}$: by iterating the argument, we see that it suffices to prove that for some k sufficiently large, any operator with symbol $r \in S^{-k}$ is continuous in L^2.

c) Let us take $k = n+1$ (in $\mathbb{R}^n$): the kernel $K(x,y)$ of $a(x,D)$ is then a continuous and bounded function, since following (3.2.2),

$$|K(x,y)| \leq (2\pi)^{-n} \int |a(x,\xi)| \mathrm{d}\xi.$$

Moreover, $(x_j - y_j)K(x,y)$ is the kernel of $\mathrm{i}\frac{\partial a}{\partial \xi_j}(x,D)$, which is 'even better' than a; iterating $n+1$ times, we finally find $(1+|x-y|^{n+1})|K(x,y)| \leq$ Const. The decrease in K at infinity implies, in particular,

$$\int |K(x,y)|\mathrm{d}y \leq C, \quad \int |K(x,y)|\mathrm{d}x \leq C. \tag{5.1.1}$$

d) It is a classical result that an operator whose kernel is continuous and satisfies (5.1.1) is bounded on L^2 (with norm at most C): indeed,

$$\begin{aligned}|Ku(x)|^2 &\leq \int |K(x,y)||u(y)|^2 \mathrm{d}y \int |K(x,y)|\mathrm{d}y \\ &\leq C\int |K(x,y)||u(y)|^2 \mathrm{d}y\end{aligned}$$

whence

$$\int |Ku(x)|^2 \mathrm{d}x \leq C\int |u(y)|^2 \mathrm{d}y \int |K(x,y)|\mathrm{d}x \leq C^2 \int |u(y)|^2 \mathrm{d}y.$$

This completes the proof of the theorem. □

5.2. Action on Sobolev spaces. We first recall the definition of these spaces.

Definition. *For $s \in \mathbb{R}$,*

$$H^s = H^s(\mathbb{R}^n) = \left\{u \in \mathcal{S}', \int (1+|\xi|^2)^s |\hat{u}(\xi)|^2 \mathrm{d}\xi < +\infty\right\}.$$

If one prefers, the operator $(1+|D|^2)^{s/2}$ is an isometry of H^s on L^2. The norm will be denoted by $|\cdot|_s$, with

$$|u|_s^2 = (2\pi)^{-n} \int (1+|\xi|^2)^s |\hat{u}(\xi)|^2 \mathrm{d}\xi.$$

When $s \in \mathbb{N}$, it is easy to see (Exercise 5.6) that

$$H^s = \{u \in L^2, \forall \alpha, |\alpha| \leq s, \partial^\alpha u \in L^2\},$$

and of course $H^0 = L^2$.

The reason why these spaces play such a big role is that they constitute a family constructed on L^2 which can be used to measure the regularity of the functions (like the C^α Hölder classes, which are constructed on L^∞).

Theorem 5.1 and Theorem 4.1 immediately give the following proposition.

Proposition 5.2. *If $a \in S^m$, the operator $a(x,D)$ maps H^s to H^{s-m} for all s.*

5.3. Gårding's inequality (weak version). If $a = a(D)$ has constant coefficients and $a(\xi) \geq 0$, the operator $a(D)$ is 'positive' in the sense that $\forall u \in \mathcal{S}$, $(a(D)u, u) \geq 0$.

In the general case, there exists a whole family of 'Gårding inequalities' which link together the positivity of the symbol and that of the operator, with an 'error' of varying magnitude.

The 'sharp' Gårding inequality is the following: if $a \in S^{2m+1}$ $(m \in \mathbb{R})$, $\operatorname{Re} a \geq 0$, then $\operatorname{Re}(a(x,D)u,u) \geq -C|u|^2_{H^m}$. In other words, we have the lower bound which we would have without the sign hypothesis on the symbol a if the latter were of order $2m$: for this reason, this inequality is also called the 'Gårding inequality with gain of a derivative'. Here, we shall only prove the 'weak' Gårding inequality.

Proposition 5.3. *If $a \in S^{2m}$ and, for $|\xi| \geq R$ and $c > 0$, $\operatorname{Re} a(x,\xi) \geq c(1+|\xi|^2)^m$, then for all N, there exists a constant C_N such that*

$$\operatorname{Re}(Au,u) + C_N|u|^2_{-N} \geq \frac{c}{2}|u|^2_m.$$

Proof. First we have $\operatorname{Re}(Au,u) = (\frac{A+A^*}{2}u,u)$; since $a^* = \overline{a} + S^{2m-1}$, we have $b = \frac{a+a^*}{2} = \operatorname{Re} a + d$ with $d \in S^{2m-1}$. Thus

$$\begin{aligned} e = b - \frac{3}{4}c(1+|\xi|^2)^m &\geq \frac{c}{4}(1+|\xi|^2)^m\left(1 - \frac{4|d|}{c(1+|\xi|^2)^m}\right) \\ &\geq \frac{c}{8}(1+|\xi|^2)^m \end{aligned}$$

for $|\xi|$ sufficiently large; the symbol $(1+|\xi|^2)^{-m}$ is of order zero and far from zero for $|\xi|$ large: Lemma 2.1.1 thus tells us that $f = e^{1/2}(1-\chi(\xi))$ is in S^m for $\chi \in C_0^\infty$, with $\chi = 1$ near 0, suitably chosen. We then have $f^*\#f = e + g$, $g \in S^{2m-1}$, whence $(Bu,u) \geq \frac{3}{4}c|u|^2_m + (Gu,u)$. Since $(Gu,u) = ((1+|D|^2)^{-m/2}Gu, (1+|D|^2)^{m/2}u)$, we finally obtain, using the Cauchy–Schwarz inequality,

$$(Bu,u) \geq \frac{3}{4}c|u|^2_m - \text{Const.}\, |u|_m|u|_{m-1}.$$

We observe that for all $\varepsilon > 0$, N,

$$|u|_{m-1} \leq \varepsilon|u|_m + C_{\varepsilon,N}|u|_{-N},$$

since this follows from the easy inequality

$$(1+|\xi|^2)^{m-1} \leq \varepsilon^2(1+|\xi|^2)^m + C^2_{\varepsilon,N}(1+|\xi|^2)^{-N}.$$

Using the inequality $2ab \leq \eta a^2 + \frac{1}{\eta}b^2$ with $a = |u|_m$, $b = |u|_{-N}$, we finally find $\operatorname{Re}(Au,u) = (Bu,u) \geq \frac{c}{2}|u|^2_m - C_N|u|^2_{-N}$, which is the inequality we seek. □

5.4. Inversion of elliptic operators. Here we realize the programme for constructing parametrices stated in the introduction.

Proposition 5.4. *Let $a \in S^m$. The conditions*

i) $\exists b \in S^{-m}; a(x,D)b(x,D) - \operatorname{id} \in \operatorname{Op}(S^{-\infty})$

and

ii) $\exists b \in S^{-m}$; $b(x, D)a(x, D) - \mathrm{id} \in \mathrm{Op}(S^{-\infty})$

are equivalent and imply

iii) $|a(x, \xi)| \geq c|\xi|^m$ *for some* $c > 0$ *and* $|\xi| \geq C$.

Conversely if iii) *is satisfied, there exists* $b \in S^{-m}$ *satisfying* i) *and* ii) *and any* b' *which satisfies* i) *or* ii) *is equal to* $b \bmod S^{-\infty}$. *An operator* a *of order* m *which satisfies* iii) *is said to be* elliptic.

Proof. If $b', b'' \in S^{-m}$ satisfy $AB' - \mathrm{id} \in \mathrm{Op}(S^{-\infty})$, $B''A - \mathrm{id} \in \mathrm{Op}(S^{-\infty})$, we have

$$B'' - B' = B''(\mathrm{id} - AB') + (B''A - \mathrm{id})B' \in \mathrm{Op}(S^{-\infty});$$

thus we also have

$$B'A - \mathrm{id} \in \mathrm{Op}(S^{-\infty}) \text{ and } AB'' - \mathrm{id} \in \mathrm{Op}(S^{-\infty}).$$

On the other hand, i) or (ii) implies $a(x, \xi)b(x, \xi) - 1 \in S^{-1}$, and in particular

$$1/2 \leq |a(x, \xi)||b(x, \xi)| \leq C|a(x, \xi)||\xi|^{-m}$$

for $|\xi|$ large, which is to say iii).

Conversely, if a is elliptic, we take

$$b = (1 + |\xi|^2)^{-m/2} F(a(1 + |\xi|^2)^{-m/2}),$$

where $F \in C^\infty(\mathbb{C})$, $F(z) = 1/z$ for $|z| \geq c'$. Then Lemma 2.1.1 implies $b \in S^{-m}$ and $ab = 1 + \chi(\xi)$, with $\chi(\xi) = 0$ for $|\xi| \geq C'$.

Theorem 4.1 then shows that

$$a(x, D)b(x, D) = \mathrm{id} - r(x, D), \quad r \in S^{-1}.$$

Setting, for $k \geq 0$, $b_k(x, D) = b(x, D)r(x, D)^k \in \mathrm{Op}(S^{-m-k})$, and choosing $b' \in S^{-m}$, $b' \sim \sum_{k \geq 0} b_k$ (Proposition 2.3), we find

$$\begin{aligned} AB' &= A\left(B' - \sum_{j<k} B_j\right) + AB\sum_{j<k} R^j = (\mathrm{id} - R)\sum_{j<k} R^j + \mathrm{Op}(S^{-k}) \\ &= \mathrm{id} - R^k + \mathrm{Op}(S^{-k}) = \mathrm{id} + \mathrm{OP}(S^{-k}), \end{aligned}$$

which is i).

An analogous argument allows us to construct B'' such that $B''A - \mathrm{id} \in \mathrm{Op}(S^{-\infty})$, and the remark at the start of the proof completes the proof of the theorem. □

One should be aware that the existence of an approximate 'left inverse' in the sense of $BA - \mathrm{id} \in \mathrm{Op}(S^{-\infty})$ does not mean that A is injective, but that $\ker A$ is formed from C^∞ functions and, more generally, that $Au \in C^\infty$ implies $u \in C^\infty$. For example, in $\mathbb{R}$, $A = \mathrm{d}/\mathrm{d}x$ is elliptic, and its kernel consists of just the constant functions. Similarly, the existence of an approximate 'right inverse' only allows us to solve the equation $Au = f$ approximately. However, if i) is satisfied, in order to obtain $u = Bv$ a solution of $Au = f$, it suffices to solve $v + Kv = f$, where $Kv(x) = \int k(x,y)v(y)dy$ has a kernel $k \in C^\infty$. Let us suppose, for example, that we wish to solve the equation $Au = f$, with Laplacian $A = \Delta = \partial_1^2 + \cdots + \partial_n^2$, and $f \in C^\infty(\mathbb{R}^n)$, locally near 0.

If β_r denotes the ball with centre 0 and radius r, we can solve in $L^\infty(\beta_r)$ the equation $v + RK\tilde{v} = f$, where $\tilde{v}$ is the extension of v by 0, outside β_r, and R is the restriction to β_r. Indeed, if we write $v = Gv$, $Gv = f - RK\tilde{v}$, we see that G is contractive in $L^\infty(\beta_r)$, whenever r is small enough, since

$$|RK\tilde{v}|_{L^\infty(\beta_r)} \leq \sup_{|x|\leq r, |y|\leq r} |k(x,y)||v|_{L^\infty(\beta_r)} r^n.$$

Taking $u = B\tilde{v}$, we thus have $Au = AB\tilde{v} = \tilde{v} + K\tilde{v}$ and $A(Ru) = v + RK\tilde{v} = f$; Ru is the desired local solution. As we noted in Section 1.1, Ru is indeed C^∞.

Another useful consequence of Proposition 5.4 is the existence of the following *a priori* inequality: if $a \in S^m$ is elliptic (in the sense of iii), that is to say uniformly on $\mathbb{R}^n$), for all $s \in \mathbb{R}$, $N \in \mathbb{R}$, there exist $C_s, C_{N,s}$, such that for all $u \in \mathcal{S}$,

$$|u|_{s+m} \leq C_s|Au|_s + C_{N,s}|u|_{-N}.$$

(This is obtained by applying the Gårding inequality (Proposition 5.3) to A^*A.) This inequality means that an elliptic operator controls, up to a small remainder, a number of derivatives equal to its order. (On this subject, see Exercises 5.12, 5.13, 5.14 and also 6.5, 6.6 and 6.7.)

6. Operators in an open subset of $\mathbb{R}^n$

Up to now, we have considered symbols defined for $(x,\xi) \in \mathbb{R}^n \times \mathbb{R}^n$, where the corresponding operators act in $\mathcal{S}'(\mathbb{R}^n)$.

In practice, we often have symbols defined for $x \in \Omega$, where Ω is an open subset of $\mathbb{R}^n$, and we wish to associate them with operators acting in $\mathcal{D}'(\Omega)$: this can be achieved by an ingenious system of cut-off functions, which we shall now explain.

6.1. Pseudo-local property.

Proposition 6.1. *Let $a \in S^m$ and let K be the kernel of $a(x,D)$. Then K is C^∞ for $x \neq y$. In particular, for all $u \in \mathcal{S}'$,*

$$\text{sing supp } a(x,D)u \subset \text{sing supp } u. \tag{6.1.1}$$

Proof. Let $x \neq y$, $\chi, \psi \in C_0^\infty$, $\chi = 1$ near x, $\psi = 1$ near y, $\operatorname{supp}\chi \cap \operatorname{supp}\psi = \emptyset$. The function $\tilde{K} = \chi(x)K(x,y)\psi(y)$ is the kernel of the operator $\chi a \psi$, whose symbol is of order $-\infty$ since $\chi a \psi \sim 0$ following Theorem 4.1; thus $\tilde{K} \in C^\infty$, that is to say $K \in C^\infty$ for $x \neq y$. If $x_0 \notin \operatorname{sing\,supp} u$, $\psi u \in C_0^\infty$ for $\psi \in C_0^\infty$, $\psi = 1$ near x_0, then

$$\chi A u = \chi A \psi u + \chi A(1-\psi)u,$$

and the operator $B = \chi A(1-\psi)$ has symbol $b \in S^{-\infty}$ if $\psi = 1$ near $\operatorname{supp}\chi$: it suffices to prove that $b(x,D)\mathcal{S}' \subset C^\infty$ in order to obtain (6.1.1).

Now, for all α, $x \mapsto D_x^\alpha(e^{ix\xi}b(x,\xi))$ is continuous with values in $\mathcal{S}$, and $D^\alpha(Bu) = (2\pi)^{-n}\langle \tilde{u}, D_x^\alpha(e^{ix\xi}b(x,\xi))\rangle$ (duality between $\mathcal{S}$ and $\mathcal{S}'$): thus $D^\alpha(Bu)$ is continuous and $Bu \in C^\infty$. □

The 'pseudo-local' property is (6.1.1).

6.2. Local symbols and operators in an open set.

Definition. *If $\Omega \subset \mathbb{R}^n$ is an open set, we define $S^m_{\mathrm{loc}}(\Omega \times \mathbb{R}^n)$ to be the set of $a \in C^\infty(\Omega \times \mathbb{R}^n)$ such that $\varphi a \in S^m(\mathbb{R}^n \times \mathbb{R}^n)$ for all $\varphi \in C_0^\infty(\Omega)$.*

If $a \in S^m_{\mathrm{loc}}(\Omega \times \mathbb{R}^n)$, the usual formula

$$a(x,D)u = (2\pi)^{-n}\int e^{ix\xi}a(x,\xi)\hat{u}(\xi)\mathrm{d}\xi$$

defines an operator from $\mathcal{S}'(\mathbb{R}^n)$ to $\mathcal{D}'(\Omega)$, which we can restrict to an operator $\mathcal{E}'(\Omega) \to \mathcal{D}'(\Omega)$ and $C_0^\infty(\Omega) \to C^\infty(\Omega)$.

The following proposition clarifies the relationship between such operators and the operators obtained 'by cut-off' from global operators.

Proposition 6.2. *Let $A : C_0^\infty(\Omega) \to C^\infty(\Omega)$ be a continuous linear operator such that for all $\varphi, \psi \in C_0^\infty(\Omega)$, $\varphi A \psi \in \mathrm{Op}(S^m)$. Then there exists $a' \in S^m_{\mathrm{loc}}(\Omega \times \mathbb{R}^n)$ with $A = a'(x,D) + R$, where R is an operator with kernel in $C^\infty(\Omega \times \Omega)$. The symbol a' here is determined modulo $S^{-\infty}_{\mathrm{loc}}(\Omega \times \mathbb{R}^n)$.*

Definition. *In the situation of Proposition 6.2, we say that A is a pseudo-differential operator on Ω, and the class of a' in $S^m_{\mathrm{loc}}/S^{-\infty}_{\mathrm{loc}}$ is called the symbol of A.*

Proof. a) Here we assume the existence of a sequence $\psi_j \in C_0^\infty(\Omega)$, with

i) $\forall x \in \Omega$, $\sum_k \psi_j(x) = 1$,

ii) for any compact $K \subset \Omega$, only a finite number of subsets from the $\operatorname{supp}\psi_j$ meet K.

Such a collection of functions is called a 'partition of unity' (for i)) and is said to be locally finite (for ii))' (see Exercise 6.1).

b) Let ψ_j be as in a) and denote $\psi_j A \psi_k = A_{jk} \in \mathrm{Op}(S^m)$. For $u \in C_0^\infty(\Omega)$, we write

$$Au = \sum_k A\psi_k u = \sum_{j,k} \psi_j A \psi_k u = \sum{}' A_{jk} u + \sum{}'' A_{jk} u,$$

where $\sum'$ denotes the sum over the j, k with $\operatorname{supp}\psi_j \cap \operatorname{supp}\psi_k \neq \emptyset$.

The function $a' = \sum' a_{jk} \in S^m_{\mathrm{loc}}(\Omega \times \mathbb{R}^n)$, since if $\varphi \in C_0^\infty(\Omega)$, $\varphi a' = \sum' \varphi a_{jk}$ contains only a finite number of terms in S^m.

The operator Ru defined by $\sum''$ has kernel $\sum'' \psi_j(x) K(x,y) \psi_k(y)$, if K is the kernel of A. Since K is C^∞ for $x \neq y$ (Proposition 6.1), the sum is in $C^\infty(\Omega \times \Omega)$, being a locally finite sum of C^∞ terms.

c) Formula (3.2.2′) shows that if $K \in C_0^\infty(\mathbb{R}^n \times \mathbb{R}^n)$ is the kernel of $b(x, D)$, then $b \in S^{-\infty}$. Applying this to the operator $\varphi a'(x, D)\psi$ in the case where a has kernel $C^\infty(\Omega \times \Omega)$, we find $a' \in S^{-\infty}_{\mathrm{loc}}(\Omega \times \mathbb{R}^n)$, which completes the proof. □

6.3. Properly supported operators.

Definition. *A continuous linear operator $A : C_0^\infty(\Omega) \to C^\infty(\Omega)$ is said to be properly supported if, for any compact subset $K \subset \Omega$, there exists a compact subset $K' \subset \Omega$ with*

$$\operatorname{supp} u \subset K \Rightarrow \operatorname{supp} Au \subset K' \text{ and } u = 0 \text{ on } K' \Rightarrow Au = 0 \text{ on } K.$$

The interest of this concept is that A maps C_0^∞ to C_0^∞, and thus A^* can be extended to an operator from $\mathcal{D}'(\Omega)$ to itself: thus we no longer need to be concerned about conditions on the support of the functions.

Moreover, the following proposition shows that 'modulo C^∞' one can always make an operator properly supported.

Proposition 6.3. *Let $A = a(x, D)$, where $a \in S^m_{\mathrm{loc}}(\Omega \times \mathbb{R}^n)$. There exists an operator R with kernel in $C^\infty(\Omega \times \Omega)$ such that $A + R$ is properly supported.*

Proof. This is the same as for Proposition 6.2, since the sum $\sum' \psi_j A \psi_k$ defines a properly supported operator. □

However, one should be aware that the notions of 'local symbol' and 'properly supported operators' do not permit any control over Au when one approaches the boundary of Ω: if one is interested in the behaviour of u right up to the boundary, possibly with boundary conditions, one has to make a special hypothesis (so-called transmission hypothesis) about the operator which makes it possible to preserve the regularity of the functions right up to the boundary (see Exercise 6.8).

7. Operators on a manifold

Let M be a C^∞ manifold and A a continuous linear operator, $A : C_0^\infty(M) \to C^\infty(M)$.

A will be said to be a pseudo-differential operator if its expression in any coordinate system is of the form $a(x, D)$ for some local symbol a. For such a definition to be usable, we must first examine the question of changes of local coordinates.

7.1. Pseudo-differential operators and changes of variables.

Proposition 7.1. *Let $\chi : \Omega \to \Omega'$ be a (C^∞) diffeomorphism between two open subsets of $\mathbb{R}^n$. Let us suppose that, for a symbol $a \in S^m$, the operator $a(x, D)$ has kernel with compact support in $\Omega \times \Omega$.*

Then

i) *the function $a'(y, \eta)$ defined by $a'(\chi(x), \eta) = e^{-i\chi(x)\eta} a(x, D) e^{i\chi(x)\eta}$ ($a' = 0$ for $y \notin \Omega'$) is a symbol of S^m,*

ii) *the kernel of $a'(x, D)$ has compact support in $\Omega' \times \Omega'$,*

iii) *for $u \in \mathcal{S}'$, $a(x, D)(u \circ \chi) = (a'(x, D)u) \circ \chi$.*

Moreover,

$$a'(\chi(x), \eta) \sim \sum \frac{1}{\alpha!} \partial_\xi^\alpha a(x^t, \chi'(x)\eta) D_y^\alpha (e^{i\rho_x(y)\eta})|_{y=x},$$

where $\rho_x(y) = \chi(y) - \chi(x) - \chi'(x)(y - x)$.

We remark that $a(x, D)(u \circ \chi)$ is well defined since the kernel of a is of compact support in $\Omega \times \Omega$ (although $u \circ \chi$ is not defined ...).

Proof. a) We first remark that if $a \in S^m$, $a(x, D)e^{ix\xi} = e^{ix\xi} a(x, \xi)$. Indeed, if $\hat{u} \in C_{\mathsf{C}}^\infty$,

$$a(x, D)u(\varepsilon x)e^{ix\xi} = e^{ix\xi}(2\pi)^{-n} \int e^{i\varepsilon x \mathcal{S}} a(x, \xi + \varepsilon \mathcal{S}) \hat{u}(\mathcal{S}) d\mathcal{S},$$

which tends to

$$e^{ix\xi} a(x, \xi) \text{ in } \mathcal{S}'(\mathbb{R}^n) \text{ if } (2\pi)^{-n} \int \hat{u}(\mathcal{S}) d\mathcal{S} \equiv u(0) = 1;$$

moreover, $u(\varepsilon x)e^{ix\xi} \to e^{ix\xi}$ in $\mathcal{S}'$, when $\varepsilon \to 0$.

b) Suppose that i) is proved. The defining formula chosen for a' means that, according to a), iii) is valid when $u(x) = e^{ix\eta}$. Since linear combinations of exponentials are dense in $\mathcal{S}'$ because of the Fourier inversion formula, we obtain iii) in general,

c) Let us prove i). If $\varphi \in C_0^\infty(\Omega)$ is such that $\varphi(x) = \varphi(y) = 1$ for (x, y) in a neighbourhood of the support of the kernel of a, we have

$$a'(\chi(x), \eta) = \varphi(x) e^{-\mathrm{i}\chi(x)\eta} a(x, D)(\varphi(x) e^{\mathrm{i}\chi(x)\eta}).$$

Setting $\lambda = 1 + |\eta|$, Theorem 3 of the Appendix gives an asymptotic expansion as a function of λ, for $a'(\chi(x), \eta)$ and all its derivatives with respect to x and η. It is also easy to convince oneself that this expansion is uniform in the parameter $\alpha = \eta/\lambda$. We surmise from these expansions that a' is a symbol in (y, η), satisfying the expansion given in iii) in the sense of symbols. □

We are now in a position to formulate a precise definition.

Definition. *The operator $A : C_0^\infty(M) \to C^\infty(M)$ will be said to be pseudo-differential of order m, if for any coordinate system $\kappa : V \to \tilde{V} \subset \mathbb{R}^n$, the transported operator $\tilde{A} : u \to [A(u \circ \kappa)] \circ \kappa^{-1}$ from $C_0^\infty(\tilde{V})$ to $C^\infty(\tilde{V})$ is pseudo-differential of order m in $\tilde{V}$, that is to say $\forall \varphi, \psi \in C_0^\infty(\tilde{V})$, $\varphi \tilde{A} \psi \in$* $\mathrm{Op}(S^m)$ *(see Proposition 6.2). We then write $A \in \Psi^m(M)$.*

Proposition 7.1 makes it possible to use this definition in practice. The definition of properly supported operators in Section 6.3 can be extended unchanged.

7.2. Principal symbol and cotangent bundle. Proposition 7.1 and the associated definition of pseudo-differential operators on M leave aside the question of the definition of a symbol for these operators; indeed, in each coordinate system, the transported operator does have a symbol determined modulo $S^{-\infty}$, but that depends on the coordinate system chosen. The question of the symbol then becomes: does there exist a function defined intrinsically, whose expression in a local coordinate system is just the symbol desired?

We shall see that the answer is yes, but under the severe restriction that the symbols should only coincide up to a symbol of order $m - 1$: only the principal symbol may thus be defined intrinsically.

7.2.1. *The cotangent bundle T^*M (review).* We recall that T^*M (the bundle of differential 1-forms on M) is the set of points $(m, \omega) \in T^*M$ consisting of a point $m \in M$ and a linear form ω on the tangent space to M at m, denoted by T_mM. The projection $\pi : T^*M \to M$ is $\pi(m, \omega) = m$, and the fibre $\pi^{-1}(m)$ is the dual space of T_mM.

a) If $(x_1, \dots, x_n)$ are the local coordinates on $V \subset M$, the vector fields $(\partial_1, \dots, \partial_n)$ form a basis of T_mM at any point of V and the forms $(\mathrm{d}x_1, \dots, \mathrm{d}x_n)$ form a basis for the dual space: denoting a 1-form by $\omega = \sum_{i=1}^n \xi_i \mathrm{d}x_i$, we obtain local coordinates (x,ξ) on $\pi^{-1}(V)$, x coordinates for m and ξ coordinates for ω. In another coordinate system $(x'_1,\dots,x'_n)$, $x'=\Phi(x)$, $x=\psi(x')$, the point m will of course have coordinates $x'(m)=\Phi(x(m))$, and the form $\omega=\sum\xi_i\mathrm{d}x_i$ will be written as $\sum\xi_i(\sum\frac{\partial\psi_i}{\partial x'_j}\mathrm{d}x'_j)$, that is to say

$$\xi'_j=\sum_i\frac{\partial\psi_i}{\partial x'_j}\xi_i,\ \text{or again}\ \xi'={}^t\psi'\xi.$$

In other words, the same point (m,ω) will have for expressions in the coordinate systems (x,ξ) and (x',ξ') the points $(x,{}^t\Phi'(x)\eta)$ and $(\Phi(x),\eta)$ in $\mathbb{R}^{2n}$, respectively.

b) On T^*M there exists a so-called 'canonical' 1-form α defined by

$$\alpha_{(m,\omega)}(Z)=\omega(\pi*Z),$$

where Z is a tangent vector to T^*M at (m,ω), and $\pi*Z$ is its projection onto M.

In the local coordinates (x,ξ) defined as in a), any vector Z can be written as $Z=\sum a_i\frac{\partial}{\partial x_i}+\sum b_j\frac{\partial}{\partial\xi_j}$,

$$\pi*Z=\sum a_i\frac{\partial}{\partial x_i},$$

and

$$\alpha_{(x,\xi)}(Z)=\left(\sum\xi_i\mathrm{d}x_i\right)(\pi*Z)=\sum\xi_ia_i=\left(\sum\xi_i\mathrm{d}x_i\right)(Z).$$

Thus

$$\alpha_{(x,\xi)}=\sum\xi_i\mathrm{d}x_i$$

where the ξ_i are the coefficients of α, which are functions on T^*M.

This 1-form plays a major role in all sorts of questions, notably through its differential $\sigma=-\mathrm{d}\alpha$, which can be written in local coordinates as $\sigma=\sum\mathrm{d}x_i\wedge\mathrm{d}\xi_i$, which means that $\sigma_{(m,\omega)}(Z,Z')=A\cdot B'-B\cdot A'$,

$$Z=A\cdot\frac{\partial}{\partial x}+B\cdot\frac{\partial}{\partial\xi},\quad Z'=A'\cdot\frac{\partial}{\partial x}+B'\cdot\frac{\partial}{\partial\xi}.$$

Thus there exists a non-degenerate 2-form on T^*M, called a 'symplectic form'.

c) This 2-form provides a bijection between the 1-forms and the vector fields on T^*M via the formula

$$\sigma(Z,X)=\omega(X)\quad(\omega\ \text{1-form};\ X,Z\ \text{vector fields}).$$

In particular, if $\omega=\mathrm{d}f$ for a function f on T^*M, the corresponding vector field Z is called the *Hamiltonian vector field of* f, denoted by H_f. In the coordinates (x, ξ), denoting

$$X = A' \cdot \frac{\partial}{\partial x} + B' \cdot \frac{\partial}{\partial \xi}, \quad \omega(X) = Xf = A' \cdot f'_x + B' \cdot f'_\xi,$$

$$\sigma(Z, X) = AB' - BA'$$

gives

$$A = f'_\xi, \quad B = -f'_x,$$

or

$$Z = H_f = f'_\xi \cdot \frac{\partial}{\partial x} - f'_x \cdot \frac{\partial}{\partial \xi}.$$

Finally, we define the *Poisson bracket* of two functions f, g on T^*M by

$$\{f, g\} = H_f g = f'_\xi \cdot g'_x - f'_x \cdot g'_\xi.$$

7.2.2. *The principal symbol.* Let us return to the situation of Proposition 7.1: if a is written as $a = a_m + S^{m-1}$, where a_m is homogeneous of degree m, the same is true for a', and

$$a'_m(\chi(x), \eta) = a_m(x, {}^t\chi'(x)\eta).$$

We shall say in this case that a has a principal symbol, which is a_m.

If, in any local map, the representative of $A \in \Psi^m(M)$ has a principal symbol, the above formula and point a) of Section 7.2.1 show that these different principal symbols are the expressions in local coordinates for a unique function on T^*M, which we call the principal symbol of A.

The impossibility of defining, for example, a symbol of order $m-1$ for A restricts the use of the symbolic calculus for operators on a manifold to the following theorem.

Theorem 7.2.2. a) *If $A_i \in \Psi^{m_i}(M)$ $(i = 1, 2)$ are properly supported and have principal symbols a_i $(i = 1, 2)$, then $A = A_1 A_2 \in \Psi^{m_1+m_2}(M)$ is properly supported and has principal symbol $a_1 a_2$.*

b) *The commutator $[A_1, A_2]$ has principal symbol $\{a_1, a_2\}$.*

Remark. We also have a theorem concerning the adjoint (see Exercise 7.5).

Concepts and assertions which are associated solely with the principal symbol, such as, for example, the ellipticity and the invertibility of elliptic operators (Proposition 5.4) remain unchanged for operators of $\Psi^m(M)$.

8. Appendix

8.1. Oscillatory integrals. We shall give a meaning to a large class of integrals of the type $\int_{\mathbb{R}^N} e^{\mathrm{i}\varphi(\theta)}a(\theta)\mathrm{d}\theta$, where φ is a rapidly varying function at infinity, and where a is regular, with essentially polynomial growth. The strong oscillation of the term $e^{\mathrm{i}\varphi(\theta)}$ when θ tends to infinity makes it possible to compensate the increase in a and to define the integral.

8.1.1. Let us begin with a crucial lemma (non-stationary phase lemma):

Lemma 1. *Let K be a compact subset of $\mathbb{R}^N$ and $\varphi \in C^\infty(\mathbb{R}^N)$ real, satisfying $|\varphi'(\theta)| \geq c_0 \geq 0$ on K. Then for any function $a \in C_0^\infty(K)$, for any $k \in \mathbb{N}$, we have*

$$\forall \lambda \geq 1,\ \lambda^k \left| \int e^{i\lambda\varphi(\theta)} a(\theta)\mathrm{d}\theta \right| \leq C_{k+1}(\varphi)C(c_0, K) \sup_{|\alpha|\leq k} |\partial^\alpha a|,$$

where $C_{k+1}(\varphi)$ remains bounded when φ remains bounded in $C^{k+1}(K)$.

Proof. Setting $L = -\mathrm{i}|\varphi'|^{-2}\sum_{j=1}^N \frac{\partial\varphi}{\partial\theta_j}\frac{\partial}{\partial\theta_j}$, we have $L(e^{\mathrm{i}\lambda\varphi}) = \lambda e^{\mathrm{i}\lambda\varphi}$, whence $\lambda^k \int e^{\mathrm{i}\lambda\varphi} a \mathrm{d}\theta = \int e^{\mathrm{i}\lambda\varphi}({}^tL)^k(a)\mathrm{d}\theta$ and we bound the last integral above by $\mathrm{vol}(K)\sup|({}^tL)^k a|$. □

8.1.2. We can now define the classes of amplitudes $a = a(\theta)$ which will arise in oscillatory integrals:

Definition. *For $\rho \in]-\infty, 1]$, $m \in \mathbb{R}$, we let $A_\rho^m(\mathbb{R}^N)$ denote the space of functions $a \in C^\infty(\mathbb{R}^N)$ satisfying*

$$\forall\theta \in \mathbb{R}^N, \forall\alpha \in \mathbb{N}^N,$$
$$|\partial^\alpha a(\theta)| \leq C_\alpha(1 + |\theta|)^{m-\rho|\alpha|} \text{ and } A_\rho^{+\infty} = \bigcup_m A_\rho^m.$$

Examples.

a) $\rho = 1$. We return to the frequency dependence of a symbol of the class S^m, introduced in Section 2.1. In particular, A_1^m contains the homogeneous functions of degree $\leq m$ (possibly cut off near 0), the polynomials of degree $\leq m$ if m is in $\mathbb{N}$, etc.

b) If $p = p(\theta)$ is homogeneous of degree $1 - \rho$ and C^∞ away from the origin, then $e^{\mathrm{i}p}$ (possibly cut off near 0) defines an element of A_p^0. In particular, $\forall\omega \in \mathbb{R}^N \setminus \{0\}$, $e^{\mathrm{i}\omega\theta}$ defines an element of A_0^0 if q is a quadratic form $e^{\mathrm{i}q} \in A_{-1}^0$, etc.

As for S^m, we define a natural structure of a complete space on A_ρ^m ($\rho \leq 1$) using semi-norms:

$$N_{\rho,k}^m(a) = \sup_{|\alpha|\leq k, \theta\in\mathbb{R}^N} (1 + |\theta|)^{-m+\rho|\alpha|}|\partial^\alpha a(\theta)|.$$

The proof of Lemma 2.2.1 can then be adapted to prove that $\mathcal{S}$ (which is just $\bigcap_{m\in\mathbb{R}} A_\rho^m$) is dense in A_ρ^m for the topology of the $A_\rho^{m+\varepsilon}, \forall\varepsilon > 0, \forall\rho \leq 1$.

8.1.3. We can now introduce precisely the integrals we wish to study.

Let $\varphi \in C^\infty(\mathbb{R}^N \setminus (0))$ be real valued and homogeneous of degree $\mu > 0$. We denote

$$I_\varphi(a) = \int_{\mathbb{R}^N} e^{\mathrm{i}\varphi(\theta)} a(\theta)\mathrm{d}\theta,$$

which is well defined for all $a \in A_\rho^m$ if $m < -N$. We wish to extend the linear form I_φ to other a. For this, we force the oscillation of the term $e^{\mathrm{i}\varphi}$ by imposing

$$d_\theta\varphi \neq 0 \text{ for } \theta \neq 0.$$

Theorem 1. *Under the above hypotheses, and if $\mu > 1 - \rho$, I_φ can be extended by continuity to all the A_ρ^m, $m \in \mathbb{R}$. This extension is unique, taking into account the density of $\mathcal{S}$ in these spaces.*

Proof. We introduce a dyadic partition of unity of $\mathbb{R}^N$ (also used in Section II.A.1.1). We write

$$1 = \chi_0(\theta) + \sum_{p=0}^{\infty} \chi(2^{-p}\theta) \text{ where } \chi_0 \in C_0^\infty, \chi \in C_0^\infty,$$

and

$$\operatorname{supp}\chi_0 \subset \{|\theta| \leq 1\}, \quad \operatorname{supp}\chi \subset \{1/2 \leq |\theta| \leq 2\};$$

then

$$I_\varphi(a) = \int_{\mathbb{R}^N} e^{\mathrm{i}\varphi}\chi_0 a + \sum_{p=0}^{\infty} \int e^{\mathrm{i}\varphi(\theta)}\chi(2^{-p}\theta)a(\theta)\mathrm{d}\theta.$$

It is a matter of proving the convergence of this series for $a \in A_\rho^m$ and of estimating its sum using a semi-norm of a in A_ρ^m. After a change of variable $\theta \mapsto 2^p\theta$, the general term of the series is estimated using Lemma 1:

$$\left|2^{Np}\int e^{\mathrm{i}2^{p\mu}\varphi(\theta)}\chi(\theta)a(2^p\theta)\mathrm{d}\theta\right|$$
$$\leq C_{k+1}2^{Np-p\mu k} \sup_{|\alpha|\leq k, 1/2\leq|\theta|\leq 2} 2^{p|\alpha|}|\partial^\alpha a(2^p\theta)|$$

for all $k \in \mathbb{N}$. We control the second term by

$$C_{k+1}2^{p(N-\mu k+m+(1-\rho)k)}N_{\rho,k}^m(a).$$

It then suffices to choose k such that

$$N + m - k(\mu - 1 + \rho) < 0,$$

to ensure the geometric convergence and obtain the desired estimate. □

Remarks.

a) The condition $\mu > 1-\rho$ is precisely that which is sufficient to ensure that the oscillating term $e^{i\varphi}$ is not in the same space of amplitudes $A_\rho^{+\infty}$ as a) (see example b)).

b) To calculate $I_\varphi(a)$, we can, for example, choose χ in $\mathcal{S}$ with $\chi(0)=1$. Then, for all $a \in A_\rho^m$, $\chi(\varepsilon\theta)a(\theta)$ tends to $a(\theta)$ in $A_\rho^{m+\delta}$ for all $\delta > 0$, when ε tends to 0. (This is Lemma 2.2.1 in its A_ρ^m version.) It follows that

$$I_\varphi(a) = \lim_{\varepsilon \to 0+} \int e^{i\varphi(\theta)} a(\theta)\chi(\varepsilon\theta)\mathrm{d}\theta.$$

This type of formula allows us to extend the usual rules of calculus for absolutely convergent integrals (Fubini's theorem, integration by parts, homogeneous change of variables, differentiation under the integral sign) to the oscillating integrals $I_\varphi(a)$.

8.2. Proof of the theorems of the symbolic calculus. Let us prove Theorem 3.2.3.

Theorem 2. *If $a = a(x,\xi) \in S^m$, then*

$$a^*(x,\xi) = \int e^{-iy\eta}\overline{a}(x-y,\xi-\eta)\mathrm{d}y\mathrm{d}\eta/(2\pi)^n$$

belongs to S^m and satisfies the asymptotic formula

$$a^*(x,\xi) \sim \sum_{\alpha \geq 0} \frac{1}{\alpha!}\partial_\xi^\alpha D_x^\alpha \overline{a}(x,\xi).$$

Proof. We set $\varphi(y,\eta) = -y\eta$: φ is a non-degenerate quadratic form on $\mathbb{R}^{2n}$ and thus satisfies the hypotheses of Theorem 1 with $\mu = 2$.

Moreover, for (x,ξ) fixed, the function $\overline{a}(x-y,\xi-\eta)$ belongs to $A_0^{m_+}(\mathbb{R}^{2n})$ (where $m_+ = \max(m,0)$), since

$$(1+|\eta|)^m \leq (1+|y|+|\eta|)^{m_+}.$$

Theorem 1 thus enables us to define the integral

$$\frac{1}{(2\pi)^n}\int e^{-iy\eta}\overline{a}(x-y,\xi-\eta)\mathrm{d}y\mathrm{d}\eta,$$

as an oscillating integral; by continuity, this quantity can only be $a^*(x,\xi)$ (defined in Section 3 as a Fourier transform). Theorem 1 immediately gives us an estimate

$$|a^*(x,\xi)| \leq CN_{0,k}^{m_+}(\overline{a}(x-\cdot,\xi-\cdot)),$$

for some k.

We shall see that this estimate gives the result if $m \geq 0$. When $m < 0$, we shall use an apparently coarser estimate, which in fact uses more information about a. In 'seeing' $\overline{a}(x-\cdot,\xi-\cdot)$ in a 'larger space' A_0^μ where $\mu \geq m_+$, we also have

$$|a^*(x,\xi)| \leq CN^\mu_{0,j}(\overline{a}(x-\cdot,\xi-\cdot)),$$

where j depends on μ.

For $\mu \geq 0$ we have

$$\begin{aligned}(1+|y|+|\eta|)^{-\mu} &\sup_{|\alpha|+|\beta|\leq k} |\partial_y^\alpha\partial_\eta^\beta \overline{a}(x-y,\xi-\eta)| \\ &\leq C(1+|\eta|)^{-\mu}(1+|\xi-\eta|)^m,\end{aligned}$$

and the choice of μ is dictated by the following elementary lemma:

Lemma 2 (Peetre's inequality). *For all $\xi,\eta\in\mathbb{R}^n$, $m\in\mathbb{R}$*

$$(1+|\xi-\eta|)^m \leq (1+|\eta|)^{|m|}(1+|\xi|)^m.$$

Proof. It is trivial if $m\geq 0$ since $1+|\xi-\eta|\leq 1+|\xi|+|\eta|\leq(1+|\xi|)(1+|\eta|)$, and we obtain the case $m<0$ by inverting the roles of ξ and $\xi-\eta$, m and $-m$. □

Thus, we choose $\mu=|m|$, and we obtain

$$|a^*(x,\xi)|\leq C(1+|\xi|)^m.$$

Applying this proof to $\partial_x^\alpha\partial_\xi^\beta a(x,\xi)$ instead of a, we deduce that $a^*\in S^m$.

For the asymptotic expansion, we apply Taylor's formula with integral remainder to the function $g(t)=\overline{a}(x+ty,\xi+t\eta)$. Remarking that

$$g^{(k)}(t)=\sum_{|\alpha|+|\beta|=k}\frac{k!}{\alpha!\beta!}y^\alpha\eta^\beta\partial_x^\alpha\partial_\xi^\beta\overline{a}(x+ty,\xi+t\eta),$$

we have

$$\begin{aligned}a^*(x,\xi) &= (2\pi)^{-n}\sum_{|\alpha|+|\beta|\leq 2k+1}\frac{(-1)^{|\alpha|+|\beta|}}{\alpha!\beta!} \\ &\quad\times\left(\int e^{-\mathrm{i}y\eta}y^\alpha\eta^\beta\mathrm{d}y\mathrm{d}\eta\right)\partial_x^\alpha\partial_\xi^\beta\overline{a}(x,\xi)+R_k(x,\xi),\end{aligned}$$

where

$$\begin{aligned}R_k(x,\xi) &= (2\pi)^{-n}\int_0^1(1-t)^{2k+1}\mathrm{d}t \\ &\quad\times\int e^{-\mathrm{i}y\eta}\sum_{|\alpha|+|\beta|=2k+2}(-1)^{|\alpha|+|\beta|} \\ &\quad\times\frac{2k+2}{\alpha!\beta!}\partial_y^\alpha\partial_\eta^\beta\overline{a}(x-ty,\xi-t\eta)y^\alpha\eta^\beta\mathrm{d}y\mathrm{d}\eta.\end{aligned}$$

Let us first calculate each term of the main part of this expansion:

Lemma 3. $\forall \alpha \in \mathbb{N}^n$, $\forall \beta \in \mathbb{N}^n$,

$$(2\pi)^{-n} \int e^{-\mathrm{i}y\eta} y^\alpha \eta^\beta \mathrm{d}y\mathrm{d}\eta = (-\mathrm{i})^{|\alpha|} \alpha! \delta_{\alpha\beta}$$

where $\delta_{\alpha\beta} = 1$ *if* $\alpha = \beta$, $\delta_{\alpha\beta} = 0$ *if* $\alpha \neq \beta$.

Proof. Let us suppose, for example, that $\alpha| \geq |\beta|$. We remark that $y^\alpha e^{-\mathrm{i}y\eta} = (-D_\eta)^\alpha(e^{-\mathrm{i}y\eta})$ and we integrate by parts:

$$(2\pi)^{-n} \int e^{-\mathrm{i}y\eta} y^\alpha \eta^\beta \mathrm{d}y\mathrm{d}\eta = (2\pi)^{-n} \int D_\eta^\alpha(\eta^\beta) e^{-\mathrm{i}y\eta} \mathrm{d}y\mathrm{d}\eta.$$

If $\alpha \neq \beta$, then there exists j such that $\alpha_j > \beta_j$. We deduce that

$$D_\eta^\alpha(\eta^\beta) = 0 \text{ if } \alpha = \beta; \text{ then } D_\eta^\alpha(\eta^\alpha) = (-\mathrm{i})^{|\alpha|}\alpha!.$$

It remains to calculate

$$(2\pi)^{-n} \int e^{-\mathrm{i}y\eta} \mathrm{d}y\mathrm{d}\eta = \lim_{\varepsilon \to 0^+} (2\pi)^{-n} \int e^{-\mathrm{i}y\eta} \chi(\varepsilon y)\chi(\varepsilon\eta) \mathrm{d}y\mathrm{d}\eta,$$

for $\chi \in \mathcal{S}(\mathbb{R}^n)$, $\chi(0) = 1$. The integral to be evaluated is just

$$(2\pi)^{-n} \int \chi(\varepsilon\eta)\hat{\chi}(\eta/\varepsilon)\mathrm{d}\eta/\varepsilon^n = (2\pi)^{-n} \int \chi(\varepsilon^2\eta)\hat{\chi}(\eta)\mathrm{d}\eta,$$

which tends to $(2\pi)^{-n} \int \hat{\chi}(\eta)\mathrm{d}\eta = \chi(0) = 1$.

We deduce that

$$a^*(x,\xi) = \sum_{|\alpha| \leq k} \frac{1}{\alpha!} \partial_\xi^\alpha D_x^\alpha \overline{a}(x,\xi) + R_k(x,\xi).$$

We note that $\partial_\xi^\alpha D_x^\alpha \overline{a} \in S^{n-|\alpha|}$. We shall prove that $R_k \in S^{m-k-1}$. For this, we remark that, integrating by parts in η and then in y as above, $R_k(x,\xi)$ is a linear combination of terms of the type

$$\int_0^1 (1-t)^{2k+1} t^j \int e^{-\mathrm{i}y\eta} \partial_y^\gamma \partial_\eta^\gamma \overline{a}(x - ty, \xi - t\eta) \mathrm{d}y\mathrm{d}\eta\mathrm{d}t,$$

where $|\gamma| \geq k+1$. But $(y,\eta) \mapsto \partial_y^\gamma \partial_\eta^\gamma \overline{a}(x - ty, \xi - t\eta)$ belongs to

$$A_0^{m-|\gamma|}(\mathbb{R}^{2n}) \subset A_0^{|m-|\gamma||}(\mathbb{R}^{2n}),$$

and the corresponding oscillating integral is thus bounded above by

$$\begin{aligned} &C \sup_{y,\eta} (1 + |y| + |\eta|)^{-|m-|\gamma||} (1 + |\xi - t\eta|)^{m-|\gamma|} \\ &\leq C \sup_\eta (1 + t|\eta|)^{-|m-|\gamma||} (1 + |\xi - t\eta|)^{m-|\gamma|} \\ &\leq C(1 + |\xi|)^{m-|\gamma|} \leq C(1 + |\xi|)^{m-k-1}. \end{aligned}$$

Thus $|R_k(x,\xi)| \leq C(1 + |\xi|)^{m-k-1}$, and we similarly find an upper bound for $\partial_x^\alpha \partial_\eta^\beta R_k(x,\xi)$ by replacing a by $\partial_x^\alpha \partial_\xi^\beta$. This completes the proof. □

Remark. A slightly different proof, which is adapted to larger classes of symbols, is proposed in Exercise 4.12 part c).

The proof of Theorem 4.1 follows exactly the same scheme, since the quantity to be studied this time is

$$a\#b(x,\xi) = \frac{1}{(2\pi)^n}\int e^{-\mathrm{i}y\eta}a(x,\xi-\eta)b(x-y,\eta)\mathrm{d}y\mathrm{d}\eta,$$

which is again an oscillating integral with amplitude in $A_0^{+\infty}(\mathbb{R}^{2n})$. We leave the details to the reader (see also Exercise 4.11).

8.3. Action of a pseudo-differential operator on an oscillating function. Using the same method, we shall prove a very useful result, which is, in particular, the key to Proposition 7.1 on change of variables for pseudo-differential operators.

Theorem 3. *Let $\psi \in C^\infty(\mathbb{R}^N)$ be real valued and such that $\forall x \in \mathbb{R}^n$, $\mathrm{d}\psi(x) \neq 0$, $u \in C_0^\infty(\mathbb{R}^n)$ and $a \in S^m(\mathbb{R}^n \times \mathbb{R}^n)$. Then for all $\lambda \geq 1$,*

$$a(x,D)(ue^{\mathrm{i}\lambda\psi})(x) = e^{\mathrm{i}\lambda\psi(x)}I(x,\lambda) \tag{3.1}$$

where, if $\lambda \to +\infty$, $I(x,\lambda)$ has the following asymptotic expansion, locally uniformly in x:

$$I(x,\lambda) \sim \sum_{\alpha\geq 0}\frac{1}{\alpha!}\partial_y^\alpha(e^{\mathrm{i}\lambda r_x(y)}u(y))|_{y=x}D_\xi^\alpha a(x,\lambda,\mathrm{d}\psi(x)). \tag{3.2}$$

In formula (3.2), we set $r_x(y) = \psi(y) - \psi(x) - \mathrm{d}\psi(x)(y-x)$, and the term of rank α is estimated by $\lambda^{m-|\alpha|/2}$.

Proof. By definition, we have, in the sense of oscillating integrals,

$$I(x,\lambda) = (2\pi)^{-n}\int\int e^{\mathrm{i}(x-y)\cdot\xi}a(x,\xi)e^{\mathrm{i}\lambda(\psi(y)-\psi(x))}u(y)\mathrm{d}y\mathrm{d}\xi.$$

Setting $y = x+z$, $\xi = \eta + \lambda\mathrm{d}\psi(x)$, we obtain

$$\begin{aligned} I(x,\lambda) &= (2\pi)^{-n}\int\int e^{-\mathrm{i}z\cdot\eta}e^{\mathrm{i}\lambda r_x(x+z)}u(x+z) \\ &\qquad \times a(x,\eta+\lambda\mathrm{d}\psi(x))\mathrm{d}z\mathrm{d}\eta \end{aligned} \tag{3.3}$$

with the notation of the statement above.

a) Let us denote $\Phi(x,\eta,\lambda) = \int e^{-\mathrm{i}z\cdot\eta}e^{\mathrm{i}\lambda r_x(x+z)}u(x+z)\mathrm{d}z$, so that

$$I(x,z) = (2\pi)^{-n}\int a(x,\eta+\lambda\mathrm{d}\psi(x))\Phi(x,\eta,\lambda)\mathrm{d}\eta,$$

and show that, for C sufficiently large,

$$\begin{aligned} &\forall k, \quad |\Phi(x,\eta,\lambda)| \leq C_k(1+|\eta|+\lambda)^{-k} \\ &\text{outside of } \frac{\lambda}{C} \leq |\eta+\lambda\mathrm{d}\psi(x)| \leq C\lambda. \end{aligned} \tag{3.4}$$

Indeed, we write

$$-z \cdot \eta + \lambda r_x(x+z) = \lambda \left[r_z(x+z) - z \cdot \frac{\eta}{\lambda} \right] = \lambda f(z)$$

with $f'(z) - \mathrm{d}\psi(x+z) = \mathrm{d}\psi(x) - \eta/\lambda$.

If $|\eta + \lambda \mathrm{d}\psi(x)| < \lambda/C$ and C is chosen so that $|\mathrm{d}\psi(y)| \geq 2C$ for any element y in the support of u, then $|f'(z)| \geq 1/C$.

Similarly, if $|\eta + \lambda \mathrm{d}\psi(x)| > C\lambda$ and C is chosen so that $|\mathrm{d}\psi(y)| \leq C/2$ for any element y in the support of u, then $|f'(z)| \geq C/2$. The integral which defines Φ thus has a uniformly non-stationary phase f and Lemma 1 implies the estimate (3.4).

Now let us write

$$I(x,\lambda) = I_1(x,\lambda) + I_2(x,\lambda),$$

where

$$\begin{aligned} I_2(x,\lambda) &= (2\pi)^{-n} \int e^{-\mathrm{i}z\cdot\eta} e^{\mathrm{i}\lambda r_x(x+z)} u(x+z) \\ &\quad \times \chi\left(\frac{\eta}{\lambda} + \mathrm{d}\psi(x)\right) a(x, \eta + \lambda \mathrm{d}\psi(x)) \mathrm{d}z \mathrm{d}\eta \end{aligned} \tag{3.5}$$

and where $\chi \in C_0^\infty(\mathbb{R}^n)$, $\chi(\zeta) = 1$ for $1/C \leq |\zeta| \leq C$, $\chi(\zeta) = 0$ for $|\zeta| \leq 1/2C$.

Because of the estimate for Φ, I_1 is rapidly decreasing when $\lambda \to +\infty$.

b) To deal with I_2, let us denote

$$b(x,z,\eta,\lambda) = u(x+z) a(x, \eta + \lambda \mathrm{d}\psi(x)) \chi\left(\frac{\eta}{\lambda} + \mathrm{d}\psi(x)\right).$$

Then b satisfies the estimates

$$\forall\alpha, \forall\beta \quad |\partial_z^\alpha \partial_\eta^\beta b(x,z,\eta,\lambda)| \leq C_{\alpha\beta} \lambda^{m-|\beta|}$$

and has a compact support in z, uniformly in η, λ and locally uniformly in x. □

The theorem is then a consequence of the following lemma:

Lemma 4. *Suppose* $r \in C^\infty(\mathbb{R}^n)$ *satisfying* $\mathrm{d}r(0) = 0$ *and* $b \in C^\infty(\mathbb{R}_z^n \times \mathbb{R}_\eta^n)$ *depending on a parameter* $\lambda \in [1, +\infty[$ *such that for some* m,

$$\forall\alpha, \forall\beta, \quad |\partial_z^\alpha \partial_\eta^\beta b(z,\eta,\lambda)| \leq C_{\alpha\beta} \lambda^{m-|\beta|}.$$

Suppose also that b *has compact support in* z *uniformly with respect to* η, λ.

Then when $\lambda \to +\infty$, *the integral*

$$J(\lambda) = (2\pi)^{-n} \int\int e^{-\mathrm{i}z\cdot\eta} e^{\mathrm{i}\lambda r(z)} b(z,\eta,\lambda) \mathrm{d}z \mathrm{d}\eta$$

admits the asymptotic expansion

$$J(\lambda) \sim \sum_{\alpha \geq 0} \frac{1}{\alpha!} \partial_z^\alpha (e^{\mathrm{i}\lambda r(z)} D_\eta^\alpha b(z, 0, \lambda))|_{z=0},$$

where the term of rank α *is estimated by* $\lambda^{m-|\alpha|/2}$.

Proof. We proceed as for Theorem 2, using a Taylor expansion of the amplitude at $z = \eta = 0$. Thanks to Lemma 3, we immediately have

$$J(\lambda) = \sum_{|\alpha|+|\beta|<2k} \frac{1}{\alpha!} \partial_z^\alpha (e^{\mathrm{i}\lambda r} D_\eta^\alpha b(z, 0, \lambda))|_{z=0} + R_k(\lambda),$$

$$\begin{aligned} R_k(\lambda) &= (2\pi)^{-n} \int_0^1 (1-t)^{2k-1} \sum_{|\alpha|+|\beta|=2k} \int\int e^{-\mathrm{i}z\cdot n} z^\alpha \eta^\beta \\ &\quad \times \partial_z^\alpha \partial_\eta^\beta (e^{\mathrm{i}\lambda r} b)(tz, t\eta, \lambda) \mathrm{d}z \mathrm{d}\eta \mathrm{d}t. \end{aligned}$$

Let us study the growth of the 'principal' terms. In $\partial_z^\alpha (e^{\mathrm{i}\lambda r} D_\eta^\alpha b(z, 0, \lambda))$ each derivative with respect to z relating to the exponential involves $\mathrm{i}\lambda \partial_z r$, which is zero at $z = 0$, so that the contribution from $e^{\mathrm{i}\lambda r}$ in this term will be a linear combination of terms of the type

$$\lambda^q \partial_z^{\alpha_1} r \ldots \partial_z^{\alpha_q} r e^{\mathrm{i}\lambda r}$$

where $\sum_{i=1}^q |\alpha_i| \leq |\alpha|$ and $|\alpha_i| \geq 2$, whence $q \leq \frac{|\alpha|}{2}$.

Taking into account the estimates for b, we conclude that

$$|\partial_z^\alpha (e^{\mathrm{i}\lambda r} D_\eta^\alpha b(z, 0, \lambda))|_{z=0} \leq C_\alpha \lambda^{m-|\alpha|/2}.$$

We shall now also exploit the annihilation of $\mathrm{d}r$ at 0 to estimate the residue $R_k(\lambda)$.

Integrating by parts with respect to η and to z, as in the proof of Theorem 2, we reduce to estimating a quantity of the type

$$\int\int e^{-\mathrm{i}z\cdot\eta} \partial_z^\gamma (e^{\mathrm{i}\lambda r} \partial_\eta^\gamma b)(tz, t\eta, \lambda) \mathrm{d}z \mathrm{d}\eta \tag{3.6}$$

uniformly for $t \in [0, 1]$, with $|\gamma| = k$.

For λ fixed, (3.6) is an oscillating integral for the phase $-z \cdot \eta$, with amplitude in A_0^0. Unfortunately, the derivatives with respect to z relating to $e^{\mathrm{i}\lambda r}$ exhibit a large growth in λ, which compensates the decrease provided by the derivatives with respect to η relating to b and means that the seminorms of this amplitude in A_0^0 do not exhibit the desired decrease.

We shall thus retransform this integral by expanding the derivatives with respect to z. The amplitude is then a linear combination of terms of the type

$$e^{i\lambda r}\lambda^q \partial_z^{\nu_1} r \ldots \partial_z^{\nu_q} r \partial_z^{\mu} \partial_\eta^{\gamma} b$$

taken at $(tz, t\eta, \lambda)$, with $\sum_{i=1}^q |\nu_i| + |\mu| = |\gamma|$, $|\nu_i| \geq 1$.

If $|\nu_i| = 1$, $(\partial_z^{\nu_1} r)(tz) = t\psi_{\nu_i}(tz) \cdot z$, where $\psi_{\nu_i} \in C^\infty(\mathbb{R}^n)$. Integrating by parts with respect to η brings us back to amplitudes of type

$$e^{i\lambda r}\lambda^q \Phi(tz) \partial_z^{\mu} \partial_\eta^{\gamma+\delta} b(tz, t\eta, \lambda),$$

where $\Phi \in C^\infty(\mathbb{R}^n)$, $|\delta| = \#\{i, |\nu_i| = 1\}$, so that

$$|\gamma \geq \sum_{i=1}^q |\nu_i| \geq 2q - |\delta|,$$

i.e.

$$q - \frac{|\delta|}{2} \leq \frac{|\gamma|}{2}. \tag{3.7}$$

We then estimate the oscillating integral by a (fixed) semi-norm of the amplitude, and, taking into account the estimates of b, we obtain

$$|R_k(\lambda)| \leq C_k \lambda^{M+q+m-|\gamma|-|\delta|}$$

or, taking into account (3.7),

$$|R_k(\lambda)| \leq C_k \lambda^{M+m-|\gamma|/2} \leq C_k \lambda^{M+m-k/2}$$

where M is an integer independent of k. This completes the proof. □

Remark. The asymptotic expansion (3.2) is obtained classically, starting from the expression (3.5), using the more general 'stationary phase' theorem (see Exercise 7.1), which is designed for studying integrals of the type $\int e^{i\lambda f(x)} u(x) dx$ when $u \in C_0^\infty$ and f only has non-degenerate critical points. (See Section 7.7 of [H1].) The particularly simple phase structure here has stimulated us to give a direct proof.

Commentary on Chapter I

Chapter I borrows all its substance, sometimes even literally, from the corresponding chapter of Hörmander's book *The Analysis of Partial Differential Operators* (Vol. III, Section 18.1). However, we have striven to write as elementary a text as possible, which is self-contained and includes what we believe to be the 'absolute minimum' amount of theory. It is our view that such a presentation will be useful not only to students, but also to more 'applied' researchers or those in other disciplines, who may read this chapter as a 'crash course'.

Of course, numerous aspects have only been touched upon: more general classes of symbols for which there exists a reasonable symbolic calculus are introduced in Hörmander [H5] (Vol. III, Section 18.5, 'Weyl calculus'); a detailed discussion of the action of 'exotic' operators in L^2 forms the subject of Coifman and Meyer's book *Au-delà des opérateurs pseudo-différentiels* [CM]; the case of analytic symbolic operators in the complex domain is developed in Sjöstrand's *Singularités analytiques microlocales* [Sj]; and, finally, a discussion of the various existing inequalities of 'Gårding type' can be found in Hörmander [H5] (Section 18.6.7, 'Sharp Gårding'; Section 18.6.8, 'Fefferman–Phong'; Section 22.3.3, 'Melin').

Other extensions of Chapter I will be discussed with the applications in Chapter II.

Exercises for Chapter I

Exercise 1.1. Let $p = p(\xi)$ be a complex-valued polynomial function of degree $m \geq 1$ on $\mathbb{R}^n$, which is elliptic in the following sense:

(∗) If p_m is the homogeneous part of highest degree of p, then $p_m(\xi) \neq 0$ for all $\xi \in \mathbb{R}^n \setminus \{0\}$.

a) Show that one can define $E \in \mathcal{S}'(\mathbb{R}^n)$ by

$$\hat{E}(\xi) = \frac{1-\chi(\xi)}{p(\xi)},$$

where $\chi \in C_0^\infty(\mathbb{R}^n)$ is to be chosen; verify that $p(D)E = \delta + \omega$, $\omega \in \mathcal{S}$.

b) Show that, if $|\alpha|$ is sufficiently large, $x^\alpha E \in L^\infty(\mathbb{R}^n)$ and $D^\beta(x^{\alpha+\beta}E) \in L^\infty(\mathbb{R}^n)$ for all β. Deduce that E is C^∞ on $\mathbb{R}^n \setminus \{0\}$.

c) Suppose that $m \geq n+1$. Show that $D^\beta E \in L^1(\mathbb{R}^n)$ for all β such that $|\beta| \leq m-n-1$. Deduce that if $u \in L^\infty(\mathbb{R}^n)$ satisfies $p(D)u \in L^\infty(\mathbb{R}^n)$, then $\partial^\alpha u \in L^\infty(\mathbb{R}^n)$ for all α of length less than or equal to $m-n-1$.
Remark. The above estimate is of course far from optimal. In fact we have $\partial^\alpha u \in L^\infty(\mathbb{R}^n)$ for $|\alpha| \leq m-1$ and 'almost' for $|\alpha| \leq m$.

Exercise 1.2. Let $u \in L^\infty(\mathbb{R}^n)$ be such that, for all $j \in \{1, \ldots, n\}$, for any integer k, $\partial_j^k u \in L^\infty(\mathbb{R}^n)$. Prove that $\partial^\alpha u \in L^\infty(\mathbb{R}^n)$ for all $\alpha \in \mathbb{N}^n$. (Hint: consider the operator $P_k = \sum_{j=1}^n \partial_j^{2k}$.)

Exercise 2.1. a) Let $a = a(x,\xi) \in S^m(\mathbb{R}^n \times \mathbb{R}^n)$. Write $n = n_1 + n_2$ with $n_2 \geq 1$ and $\xi = (\xi_1, \xi_2)$ with $\xi_i \in \mathbb{R}^{n_i}$. Suppose that a does not depend on ξ_2. Show that a is differential of order $[m]$.

b) Suppose $a = a(x,\xi) \in C^\infty(\mathbb{R}^n \times \mathbb{R}^n)$ satisfies

$$\forall \alpha, \forall \beta, \quad |\partial_x^\alpha \partial_\xi^\beta a(x,\xi)| \leq C_{\alpha\beta}(1 + |\xi_1|)^{m-|\beta|},$$

with the notation of a). Suppose $\chi \in S^0(\mathbb{R}^n_\xi)$ has support in $\{|\xi_2| \leq C|\xi_1|\}$ for some $C > 0$. Show that $a(x,\xi)\chi(\xi) \in S^m$. (Justify the existence of such a χ.)

Exercise 2.2. Let $a = a(x,\xi) \in C^\infty(\mathbb{R}^n \times \mathbb{R}^n)$ be bounded together with its derivatives on $\mathbb{R}^n \times \{|\xi| \leq 1\}$, and $\chi \in C^\infty(\mathbb{R}^n_\xi)$, with support contained in $\{1/2 \leq |\xi| \leq 2\}$, where χ does not vanish on S^{n-1}.

For $\lambda \geq 1$, set $a_\lambda(x,\xi) = \chi(\xi)a(x,\lambda\xi)$. Show that the following conditions are equivalent:

i) $a \in S^m$.

ii) $\forall k \in \mathbb{N}$, $\|a_\lambda\|_{C^k(\mathbb{R}^n \times \mathbb{R}^n)} \leq C_k \lambda^m$.

Exercise 2.3. Suppose $a = a(x,\xi) \in C^\infty(\mathbb{R}^n \times \mathbb{R}^n)$ satisfies: $\exists m \in \mathbb{R}$

i) $\forall \alpha \in \mathbb{N}$, $|\partial_x^\alpha a(x,\xi)| \leq C_\alpha(1+|\xi|)^m$,

ii) $\forall \beta \in \mathbb{N}^n$, $|\partial_\xi^\beta a(x,\xi)| \leq C_\beta(1+|\xi|)^{m-|\beta|}$.

Show that $a \in S^m$. (Use Exercises 2.2 and 1.2).

Exercise 2.4. a) Let $f \in C^k(\mathbb{R}^n)$. Suppose that f and $\partial^\alpha f(|\alpha| = k)$ are bounded. Let $x \in \mathbb{R}^n$ and $P_x(h) = \sum_{1\leq|\beta\leq k-1} \sum \frac{\partial^\beta f(x)}{\beta!} h^\beta$. Show that the family $(P_x)_{x\in\mathbb{R}^n}$ is bounded and deduce that

$$\forall \beta, 1 \leq \beta \leq k-1,\ \|\partial^\beta f\|_{L^\infty} \leq C\left(\|f\|_{L^\infty} + \sum_{|\alpha|=k} \|\partial^\alpha f\|_{L^\infty}\right).$$

By changing x to λx (λ to be chosen), prove that

$$\|\partial^\beta f\|_{L^\infty} \leq C\|f\|_{L^\infty}^{1-|\beta|/k}\left(\sum_{|\alpha|=k} \|\partial^\alpha f\|_{L^\infty}\right)^{|\beta|/k}.$$

b) Let $p = p(x,\xi) \in S^m$ be such that there exists $\mu < m$ satisfying $|p(x,\xi)| \leq C(1+|\xi|)^\mu$. Show that, $\forall \varepsilon > 0$, $p \in S^{\mu+\varepsilon}$ (one might use Exercise 2.2).

c) To what natural question does the function $\xi \mapsto e^{\mathrm{i}(\log|\xi|)^2}$ correspond?

Exercise 2.5. Suppose $C_0 > 0$, $\Omega = \{\zeta \in \mathbb{C}^n, |\mathrm{Re}\,\zeta| > C_0|\mathrm{Im}\,\zeta|\}$ and let $a = a(\zeta)$ be a holomorphic function on Ω satisfying

$$\forall \zeta \in \Omega,\ |\zeta| \geq 1,\ |a(\zeta)| \leq C(1+|\zeta|)^m.$$

Show that $(1-\chi(\xi))a(\xi) \in S^m$ ($\chi \in C_0^\infty(\mathbb{R}^n)$ is equal to 1 near $\xi = 0$).

Exercise 2.6. This exercise extends the lemma of Section 2.3 to the case of functions holomorphic in a convex sector of $\mathbb{C}$. Let $\Omega = \{z \in \mathbb{C}, |\mathrm{Im}\,z| < C_0 \,\mathrm{Re}\,z\}$, where $C_0 > 0$ is given.

a) For $\alpha > 0$, set $f_\alpha(z) = e^{-\alpha/z}$. Verify that $f_\alpha(z)/z^k \to 0$ when $z \to 0$, $z \in \Omega$ for all $k \in \mathbb{N}$ and that $f_\alpha(z) \to 1$ when $\alpha \to 0^+$ uniformly on any compact subset of Ω.

b) Let $(b_j)_{j\in\mathbb{N}}$ be a sequence of complex numbers. Show that there exists a sequence $(\alpha_j)_{j\in\mathbb{N}}$ of positive real numbers such that the series

$$\sum_{j=0}^{\infty} b_j \frac{z^j}{j!}(1 - f_{\alpha_j}(z))$$

converges uniformly on any compact subset of Ω. Conclusion?

Exercise 2.7. Let m_j be a decreasing sequence of reals tending to $-\infty$, and let $a_j \in S^{m_j}$. Suppose that there exists $a \in C^\infty(\mathbb{R}^n \times \mathbb{R}^n)$ satisfying

i) $\exists \mu \in \mathbb{R}, a \in S^\mu$,

ii) $\exists (\mu_n)_{n\in\mathbb{N}}$, a sequence of reals tending to $-\infty$, such that

$$\forall n \in \mathbb{N}, \quad \left| a(x,\xi) - \sum_{j<n} a_j(x,\xi) \right| \le C_n(1+|\xi|)^{\mu_n}.$$

Show that $a \sim \sum a_j$. (One might use Exercise 2.4.)

Exercise 2.8. For $\tau \in \mathbb{R}$, $\xi \in \mathbb{R}$, set $p(\xi,\tau) = 1/(1+\xi^2+i\tau)$.

a) Establish a recurrence relation between $\partial_\xi^{k+2}p, \partial_\xi^{k+1}p, \partial_\xi^k p$. Deduce that p does not belong to any $S^m(\mathbb{R}^2_{\xi,\tau})$. (One might reason by contradiction and study the decrease in the $\partial_\xi^k p(\tau^{1/2},\tau)$ for $\tau \to +\infty$.)

b) Show that the following estimates are available:

$$\forall j, \forall k \quad |\partial_\xi^k \partial_\tau^j p(\xi,\tau)| \le C_{j,k}(1+|\xi|+|\tau|)^{-1-k/2-j}.$$

Exercise 2.9. For $\rho \in [0,1]$, $\delta \in [0,1]$, let $S^m_{\rho,\delta}(\mathbb{R}^n \times \mathbb{R}^n)$ denote the vector space of functions $a = a(x,\xi) \in C^\infty(\mathbb{R}^n \times \mathbb{R}^n)$ such that

$$\forall \alpha, \forall \beta, \quad |\partial_x^\alpha \partial_\xi^\beta a(x,\xi)| \le C_{\alpha\beta}(1+|\xi|)^{m-\rho|\beta|+\delta|\alpha|}.$$

a) Let $b \in C^\infty(\mathbb{R}^n \times \mathbb{R}^n)$ satisfy $|b| \ge C(1+|\xi|)^\mu$ $(C > 0)$ and suppose

$$\exists \rho, \delta, \forall \alpha, \forall \beta, \quad |\partial_x^\alpha \partial_\xi^\beta b| \le C_{\alpha\beta}|b(x,\xi)|(1+|\xi|)^{-\rho|\beta|+\delta|\alpha|}.$$

Show that $1/b \in S^{-\mu}_{\rho,\delta}$.

b) In which classes $S^m_{\rho,\delta}$ are the symbols $\varphi(x)(1+|x|^{2\mu}|\xi|^{2\nu})^{-1}$ (μ, ν integers ≥ 0, $\varphi \in C_0^\infty(\mathbb{R}^n)$)?

Exercise 3.1. This exercise establishes the type of conditions satisfied by the kernel of a pseudo-differential operator (see Section 3.2.2) without seeking to characterize the order of the operator.

a) Let $a(x,\xi) \in \bigcup_{m\in\mathbb{R}} S^m (\equiv S^{+\infty})$. Consider the kernel K of $a(x, D)$ given by formula (3.2.2). Show that there exists an integer N such that

$$(*) \qquad \begin{array}{l} \forall\alpha \in \mathbb{N}^n, \forall\beta \in \mathbb{N}^n, \\ |\beta| \geq N, \quad (\partial_x + \partial_y)^\alpha (x-y)^\beta K(x,y) \in C^{|\beta|-N}(\mathbb{R}^{2n}) \end{array}$$

(i.e. it is continuous and bounded on $\mathbb{R}^{2n}$ together with its derivatives up to order $|\beta| - N$).

b) Conversely, suppose $K \in \mathcal{S}'(\mathbb{R}^{2n})$ satisfies the conditions $(*)$ for some N. Let $a \in \mathcal{S}'(\mathbb{R}^{2n})$ denote the distribution defined by (3.2.2′). Show that there exists an integer M such that

$$\forall\alpha \in \mathbb{N}^n, \forall\beta \in \mathbb{N}^n, \quad |\beta| \geq M, \forall\gamma \in \mathbb{N}^n, |\gamma| \leq |\beta| - M, \quad \xi^\gamma \partial_x^\alpha \partial_\xi^\beta a \in L^\infty.$$

Deduce that $a \in S^{+\infty}$.

Remark. If one seeks to determine the order of a precisely from the integer N in a) and b), one finds a major discrepancy (of the order of n). This has to do with the fact that the space L^∞ behaves very badly with respect to Fourier analysis.

Exercise 3.2. Exercise 3.1 shows that the kernel of a pseudo-differential operator is C^∞ away from the diagonal $\{x = y\}$.

Let $a \in S^0$ and let K be the kernel of $a(x, D)$. Here, it is proposed to study the behaviour of $K(x, y)$ when (x, y) approaches the diagonal, i.e. when $z = x - y$ is close to 0. Set $H(x, z) = K(x, x - z)$. For $\lambda \geq 1$, to be chosen, denote $H = H_1 + H_2$ where

$$\begin{aligned} H_1(x,z) &= \int e^{\mathrm{i}z\cdot\xi} a(x,\xi)\chi(\xi/\lambda)\mathrm{d}\xi/(2\pi)^n, \\ H_2(x,z) &= \int e^{\mathrm{i}z\cdot\xi} a(x,\xi)(1-\chi(\xi/\lambda))\mathrm{d}\xi/(2\pi)^n, \end{aligned}$$

where $\chi \in C_0^\infty(\mathbb{R}^n)$ is equal to 1 for $|\xi| \leq C$.

a) By integrating by parts, show that, for all N

$$|H_2(x,z)| \leq C_N \lambda^{n-2N} |z|^{-2N}.$$

b) Verify that $|H_1(x,z)| \leq C\lambda^n$, and conclude that

$$|H(x,z)| \leq C/|z|^n.$$

c) Show, similarly, that

$$|\partial_x^\alpha \partial_y^\beta K(x,y)| \leq \frac{C_{\alpha\beta}}{|x-y|^{n+|\beta|}}$$

for all $\alpha, \beta \in \mathbb{N}^n$.

Remark. The operators with kernel K satisfying the conditions described in c) form a much more general class than that of the pseudo-differential operators of order 0. In particular, they are not, in general, bounded on L^2 (unlike the latter, see Section 5), and the description of this property as a function of the distribution K is a delicate task (see [DJ]). A simple example is studied in Exercise 5.4.

Exercise 4.1. Show that, if $a_1(x, D)$ and $a_2(x, D)$ are differential operators, then $a_1(x, D)a_2(x, D) = b(x, D)$ where $b(x, \xi) = \sum_{\alpha \geq 0} \frac{1}{\alpha!} \partial_\xi^\alpha a_1 D_x^\alpha a_2$ (note that the sum is finite).

Exercise 4.2. Let A be a nilpotent pseudo-differential operator (i.e. $\exists k \in \mathbb{N}$ such that $A^k = 0$). What can one say about A? Give a non-trivial example with $k = 2$. (One might use Exercise 2.4.)

Exercise 4.3. Let A be an idempotent pseudo-differential operator (i.e. $A^2 = A$). Assume that the dimension n is at least 2. Show that either A or $1 - A$ is of order $-\infty$.

Exercise 4.4. Without using the proof of Theorem 4.1 in the Appendix, prove the following results:

a) The mapping

$$\begin{array}{rcl} S^{m_1} \times S^{m_2} & \to & S^{m_1+m_2} \\ (a_1, a_2) & \mapsto & a_1 \# a_2 \quad \text{(symbol for } a_1(x, D)a_2(x, D)) \end{array}$$

is continuous (use the closed graph theorem and, for example, Proposition 3.1.

b) The mapping

$$\begin{array}{rcl} S^{m_1} \times S^{m_2} & \to & S^{m_1+m_2-1} \\ (a_1, a_2) & \mapsto & a_1 \# a_2 - a_1 a_2 \end{array}$$

is continuous.

Exercise 4.5. Characterize the space A_ρ^m of the Appendix when $\rho > 1$, as follows:

$$\begin{aligned} &\text{if } m < 0, A_\rho^m = 8, \\ &\text{if } m \geq 0, A_\rho^m = \mathcal{S} \oplus \mathcal{P}_{[m/\rho]}, \end{aligned}$$

where $\mathcal{P}_d$ is the space of polynomial functions of degree d on $\mathbb{R}^N$.

Exercise 4.6. a) Verify the following statement of Theorem 1 'with parameter': *Let Ω be an open subset of $\mathbb{R}^N$, and suppose $\varphi = \varphi(x, \theta) \in C^\infty(\Omega \times \mathbb{R}^N \setminus 0)$ and $a = a(x, \theta) \in C^\infty(\Omega \times \mathbb{R}^N)$ satisfy*

i) *φ is μ-homogeneous in θ, and $d_\theta \varphi(x, \theta) \neq 0$ on $\Omega \times \mathbb{R}^N \setminus 0$,*

ii) *there exists* $\rho > 1 - \mu$ *such that for any compact subset* K *of* Ω, $x \in K$, $\theta \in \mathbb{R}^N$,

$$\forall\alpha, \forall\beta, \quad |\partial_x^\alpha \partial_\theta^\beta a(x,\theta)| \le C_{\alpha,\beta,K}(1+|\theta|)^{m-\rho|\beta|}.$$

Then the function $I_\varphi(a)(x) = \int e^{i\varphi(x,\theta)} a(x,\theta) d\theta$ *is* C^∞ *on* Ω.

b) It is proposed to generalize the above result when the condition i) is relaxed to

i′) φ *is* μ*-homogeneous in* θ, *and* $d_{x,\theta}\varphi(x,\theta) \neq 0$ *on* $\Omega \times \mathbb{R}^N \setminus 0$.

We shall show that in that case $I_\varphi(a)$ can still be defined, but as a distribution on Ω, where the mapping $I_\varphi : A_\rho^{+\infty} \to \mathcal{D}'(\Omega)$ is continuous.

Suppose a satisfies ii) with $m < -N$. Then $I_\varphi(a) \in C^0(\Omega)$ and, if $u \in C_0^\infty(\Omega)$, we have

$$\int I_\varphi(a) u dx = \int\int e^{i\varphi(x,\theta)} a(x,\theta) u(x) d\theta dx.$$

Let $x_0 \notin \operatorname{supp} u$, and let U be an open neighbourhood of $\operatorname{supp} u$ which is relatively compact in Ω. Choose a function $\chi = \chi(t,\theta) \in C^\infty((\mathbb{R}^n \times \mathbb{R}^n)\setminus\{0\})$ which is homogeneous of degree 0 in (t,θ), such that $\chi = 1$ if $x_0 + \frac{t}{|\theta|} \in \operatorname{supp} u$, $\chi = 0$ if $x_0 + \frac{t}{|\theta|} \notin U$. Then set

$$\begin{aligned} \psi(t,\theta) &= \varphi\left(x_0 + \frac{t}{|\theta|}, \theta\right)\chi(t,\theta), \\ b(t,\theta) &= \frac{1}{|\theta|^N} a\left(x_0 + \frac{t}{|\theta|}, \theta\right) u\left(x_0 + \frac{t}{|\theta|}\right). \end{aligned}$$

Show that $\psi \in C^\infty(\mathbb{R}^{n+N} \setminus \{0\})$, that it is homogeneous of degree μ with respect to (t,θ), and that $d_{t,\theta}\psi \neq 0$ if $x_0 + \frac{t}{|\theta|} \in \operatorname{supp} u$. Show that $b \in A_\rho^{m-N}(\mathbb{R}^{n+N})$ and that

$$\int I_\varphi(a) u dx = \int\int e^{i\psi} b dt d\theta.$$

Conclude.

c) Show that the singular support of $I_\varphi(a)$ is contained in the set $\{x \in \Omega, \exists\theta \in \mathbb{R}^N \setminus 0, d_\theta\varphi(x,\theta) = 0\}$.

Remark. The distributions of type $I_\varphi(a)$ play a particularly important role in analysis, especially when $\mu = 1$, $\rho = 1$ (Fourier integrals). Some examples are given in Sections II.B.1 and II.B.2.

Exercise 4.7. Let a be a real number. Let S_α denote the singular support of the Fourier transform of the function f_α defined on $\mathbb{R}$ by $f_\alpha(x) = \cos|x|^\alpha$. Show that $S_\alpha = \emptyset$ if $\alpha > 1$, $S_\alpha = \{\pm 1\}$ if $\alpha = 1$, $S_\alpha \subset \{0\}$ if $\alpha < 1$. The next exercise proves that in fact $S_\alpha = \{0\}$ if $\alpha < 1$.

Exercise 4.8. Let $a = a(\theta) \in A^m_\rho$, $\rho > 0$. Suppose that $\hat{a}$ is C^∞ near 0. Show that $a \in \mathcal{S}$. (Hint: estimate $a(\theta) - (\hat{\varphi} * a)(\theta)$, where $\varphi \in C_0^\infty$, $\int \hat{\varphi} = 1$ when $\theta \to \infty$.)

Exercise 4.9. Let q be a non-degenerate quadratic form on $\mathbb{R}^n$. Exercise 4.6 shows that the Fourier transform of $e^{\mathrm{i}q/2}$ is a C^∞ function on $\mathbb{R}^n$ (why?). We shall calculate it explicitly.

a) $\hat{q}$ will be called the dual form for q, i.e., if $q(x) = \langle Ax, x\rangle$, $\hat{q}(\xi) = \langle A^{-1}\xi, \xi\rangle$.

Verify that $q(x) - 2\langle x, \xi\rangle = q(x - A^{-1}\xi) - \hat{q}(\xi)$, and deduce that

$$\widehat{e^{\mathrm{i}q/2}} = (2\pi)^{n/2} e^{\mathrm{i}\sigma\pi/4} |\det q|^{-1/2} e^{-\mathrm{i}\hat{q}/2}$$

where σ is the signature of q and $\det q$ is the determinant of q in an orthonormal basis. (Calculate $\int_{-\infty}^{+\infty} e^{\mathrm{i}\lambda x^2/2} \mathrm{d}x$ by analytic continuation starting from the Gaussian integral $\int_{-\infty}^{+\infty} e^{-\lambda x^2/2} \mathrm{d}x = \sqrt{2\pi/\lambda}$.)

b) Use a) to deduce the formula

$$\forall \xi \in \mathbb{R}^n, \forall k \in \mathbb{N}, \quad \sum_{|\alpha| = 2k} \left(\int e^{\mathrm{i}q(x)/2} \frac{x^\alpha}{\alpha!} \mathrm{d}x \right) \xi^\alpha$$
$$= (2\pi)^{n/2} e^{\mathrm{i}\sigma\pi/4} |\det q|^{-1/2} \times \frac{(\mathrm{i}\hat{q}(\xi))^k}{2^k k!}$$

which will be useful in establishing the stationary phase formula (see Exercise 7.1).

Exercise 4.10. Returning to the proof of Theorem 1, prove the following result.

Let $\varphi = \varphi(\theta)$ and $a = a(\theta, \lambda)$ be two functions satisfying the same hypotheses as in the theorem, where a varies with the parameter $\lambda \in [1, +\infty[$ as follows:

$$\text{if } \rho \geq 0, \quad |\partial^\alpha a(\theta, \lambda)| \leq C_\alpha (1 + |\theta|)^{m - \rho|\alpha|},$$
$$\text{if } \rho < 0, \quad |\partial^\alpha a(\theta, \lambda)| \leq C_\alpha (1 + |\theta| + \lambda)^{m - \rho|\alpha|}.$$

Suppose $\chi_0 \in C_0^\infty(\mathbb{R}^N)$ equals 1 near $\theta = 0$. Show that for all $N \in \mathbb{N}$,

$$\left| \int e^{\mathrm{i}\varphi(\theta)} a(\theta) a(\theta, \lambda)(1 - \chi_0(\theta/\lambda)) \mathrm{d}\theta \right| \leq C_N \lambda^{-N}.$$

Exercise 4.11. Let $p = p(x, y, \xi) \in C^\infty(\mathbb{R}^n \times \mathbb{R}^n \times \mathbb{R}^n)$. We say that $p \in S^m(\mathbb{R}^{2n} \times \mathbb{R}^n)$ if

$$|\partial_x^\alpha \partial_y^\beta \partial_\xi^\gamma p(x, y, \xi)| \leq C_{\alpha\beta\gamma} (1 + |\xi|)^{m - |\gamma|}.$$

Under this hypothesis, consider the operator A with kernel

$$K(x,y) = \frac{1}{(2\pi)^n}\int e^{\mathrm{i}(x-y)\cdot\xi} p(x,y,\xi)\mathrm{d}\xi.$$

Show that A is a pseudo-differential operator of order m and that its symbol $a(x,\xi)$ satisfies

$$a(x,\xi) \sim \sum_{\alpha\geq 0} \frac{1}{\alpha!}\partial_y^\alpha D_\xi^\alpha p(x,y,\xi)_{y=x}.$$

Verify that the above theorem implies Theorems 3.2.3 and 4.1. (For the latter, note that if $A = a(x,D)$ and $C = c(x,D)$ are two pseudo-differential operators, then AC^* is of the type studied above with $p(x,y,\xi) = a(x,\xi)\overline{c(y,\xi)}$.)

Exercise 4.12 (*). Let $\rho \in [0,1], \delta \in [0,1[$; as in Exercise 2.9, define the class $S^m_{\rho,\delta}(\mathbb{R}^{2n}\times\mathbb{R}^n)$ by $p = p(x,y,\xi) \in S^m_{\rho,\delta}(\mathbb{R}^{2n}\times\mathbb{R}^n)$ if

$$|\partial_x^\alpha\partial_y^\beta\partial_\xi^\gamma p(x,y,\xi)| \leq C_{\alpha\beta\gamma}(1+|\xi|)^{m-\rho|\gamma|+\delta(|\alpha|+|\beta|)}.$$

The operator A is defined as in Exercise 4.11.

a) Verify that the integral giving the symbol $a(x,\xi)$ for A as a function of p can also be interpreted using Theorem 1.

b) Suppose $\delta = 0$. Prove that $a \in S^m_{\rho,0}$.

c) Suppose that $\rho \geq \delta > 0$. Denote $\lambda = |\xi|$, and choose $\chi_0 \in C_0^\infty(\mathbb{R}^{2n})$ equal to 1 for $|y|+|\eta| \leq 1/2$ and equal to 0 for $|y|+|\eta| \geq 3/4$. Show using Exercise 4.10 that

$$\left|a(x,\xi) - \int e^{\mathrm{i}y\cdot\eta} p(x,x-y,\xi-\eta)\chi_0(y/\lambda,\eta/\lambda)\mathrm{d}y\mathrm{d}\eta/(2\pi)^n\right| \leq C_N\lambda^{-N},$$

for all $N \in \mathbb{N}$.

Set $y = \lambda^{-\delta}y'$, $\eta = \lambda^\delta\eta'$ in the integral compared above to $a(x,\xi)$ and verify that the function

$$q(x,y,\xi,\eta) = p(x,x-\lambda^{-\delta}y,\xi-\lambda^\delta\eta)\chi_0(\lambda^{-\delta}y/\lambda,\lambda^\delta\eta/\lambda)$$

satisfies the estimates

$$\forall\alpha,\beta,x,y,\xi,\eta \quad |\partial_y^\alpha\partial_\eta^\beta q| \leq C_{\alpha\beta}\lambda^m.$$

Conclude that $|a(x,\xi)| \leq C(1+|\xi|)^m$, since $a \in S^m_{\rho,\delta}$.

d) Suppose $\rho > \delta \geq 0$. Prove that a again satisfies

$$a(x,\xi) \sim \sum_{\alpha\geq 0} \frac{1}{\alpha!}\partial_y^\alpha D_\xi^\alpha p(x,y,\xi)_{y=x}.$$

Remark. This exercise shows, in particular, that passages to the adjoint and the product leave the class of pseudo-differential operators of type (ρ,δ)

stable when $\rho \geq \delta$, $\delta < 1$. Moreover, if $\rho > \delta$, these operators have a symbolic calculus completely analogous to that described in Sections 3 and 4. Finally we note that the assumption $\delta < 1$ is crucial: the adjoint of an operator of type (1,1) is not, in general, a pseudo-differential operator of type (1,1). (For a detailed study, see [H8].)

Exercise 5.1. Let $x_0 \in \mathbb{R}^n$, $\xi_0 \in \mathbb{R}^n$ with $|\xi_0| = 1$, $\varphi \in \mathcal{S}(\mathbb{R}^n)$, $\lambda \geq 1$. Consider $u_\lambda(x) = \lambda^{n/4} e^{i\lambda x\cdot\xi_0}\varphi(\lambda^{1/2}(x - x_0))$.

a) Calculate $|u_\lambda|_0$.

b) Let $a \in S^m$. Show that there exists $b_\lambda \in S^m$ such that

$$e^{-i\lambda x\cdot\xi_0} a(x,D)u_\lambda = \lambda^{n/4} b_\lambda(y,D)\varphi|_{y=\lambda^{1/2}(x-x_0)}.$$

Show that there exists $M \in \mathbb{N}$ such that

$$|b_\lambda(y,\eta) - a(x_0,\lambda\xi_0)| \leq C\lambda^{m-1/2}(1 + |y| + |\eta|)^M.$$

(Distinguish two classes, depending on whether $|\eta|$ is less than or greater than $\lambda^{1/2}/2$.)

c) Deduce that $(a(x,D)u_\lambda, u_\lambda) = a(x_0,\lambda\xi_0)|\varphi|_0^2 + 0(\lambda^{m-1/2})$.

d) Show that if $a(x,D)$ is bounded on L^2, then $a \in S^\varepsilon$ for all $\varepsilon > 0$ and $|a(x,\xi)| \leq C$ (use Exercise 2.4, starting by noting that $|a(x,\xi)| \leq C(1 + |\xi|)^{m-1/2}$).

Exercise 5.2. This exercise concerns the first proof (historically) of the L^2 continuity theorem for a pseudo-differential operator: the symbol is assumed to be homogeneous of degree 0.

Thus, suppose $a = a(x,\xi) \in S^0(\mathbb{R}^n \times \mathbb{R}^n)$ coincides for $|\xi| \geq 1$ with a function $\sigma(x,\xi)$ which is homogeneous of degree 0 in ξ. (Note that this degree of homogeneity enables us to define $\sigma(x,D)$.)

a) Show that if $\sigma(x,D)$ is bounded on L^2, then $a(x,D)$ is bounded on L^2. (One might study $a(x,D) - \sigma(x,D)\chi(D)$ where $\chi \in C^\infty(\mathbb{R}^n)$ is equal to 1 for $|\xi| \geq 1$ and 0 near $\xi = 0$).

b) Suppose that $n = 1$. Show that $\sigma(x,D)$ is bounded on $L^2(\mathbb{R})$.

c) Suppose that $n \geq 2$ and assume the following theorem regarding the decomposition into spherical harmonics. *There exists a Hilbert basis $(h_k)_{k\in\mathbb{N}}$ for $L^2(S^{n-1})$ formed from C^∞ functions on S^{n-1}, such that the Fourier coefficients of any function $f \in C^\infty(S^{n-1})$ are rapidly decreasing as $k \to +\infty$, while $\|h_k\|_{L^\infty(S^{n-1})} = O(k^M)$ for some M.*

For $n = 2$, this reduces to the usual Fourier series decomposition.

Show that one can write

$$\sigma(x,\xi) = \sum_{k=0}^{\infty} \sigma_k(x) h_k(\xi/|\xi|),$$

where $\|\sigma_k\|_{L^\infty(\mathbb{R}^n)}$ is rapidly decreasing in k. Deduce that $\sigma(x, D)$ is bounded on $L^2(\mathbb{R}^n)$.

Remark. Note that the above proof does not require any regularity in x for a or σ but only that they should be in L^∞. This has to do with the assumption we made on σ regarding homogeneity in ξ.

More precisely, if a only satisfies the estimates

$$\forall\alpha,\ |\partial_x^\alpha a(x,\xi)| \leq C_\alpha(1+|\xi|)^{-|\alpha|},$$

the operator $a(x, D)$ is not necessarily bounded on L^2.

On the other hand, it can be shown (see Theorem 9 of [Hw]) that it is bounded if one adds the following estimates

$$\forall i \in \{1,\dots,n\}, \forall\alpha, \quad |\partial_{x_i}\partial_\xi^\alpha a(x,\xi)| \leq C_\alpha(1+|\xi|)^{-|\alpha|}.$$

Exercise 5.3 (*). This exercise gives a proof of Theorem 5.1 which does not use the symbolic calculus.

Suppose $a \in C^{2n}(\mathbb{R}^n \times \mathbb{R}^n)$ satisfies $|\partial_x^\alpha \partial_\xi^\beta a| \leq C$ for $\alpha \in \{0,1\}^n$, $\beta \in \{0,1\}^n$. It is proposed to show that $a(x, D)$ is bounded on L^2. Set $\|a\| = \sum_{\alpha\in\{0,1\}^n, \beta\in\{0,1\}^n} \sup|\partial_x^\alpha \partial_\xi^\alpha a|$.

a) Preliminary. Let $\varphi \in L^2(\mathbb{R}^n)$, $u \in L^2(\mathbb{R}^n)$, $v \in L^2(\mathbb{R}^n)$. Verify that

$$\tilde{u}(x,\xi) = \int e^{-\mathrm{i}y\cdot\xi}\varphi(x-y)u(y)\mathrm{d}y$$

and

$$\tilde{v}(x,\xi) = \int e^{\mathrm{i}x\cdot\eta}\varphi(\xi-\eta)\hat{v}(\eta)\mathrm{d}\eta$$

are in $L^2(\mathbb{R}^{2n})$.

b) Verify that it suffices to prove the following inequality:

$$\forall a \in C_0^\infty(\mathbb{R}^n\times\mathbb{R}^n), \forall u \in \mathcal{S}, \forall v \in \mathcal{S},$$
$$|(a(x,D)u, v)| \leq C\|a\||u|_{L^2}|v|_{L^2}.$$

c) Set $\varphi(x) = \prod_{j=1}^n (1+\mathrm{i}x_j)^{-1}$ and then let $\tilde{u}$ and $\tilde{v}$ be defined as in a). Write

$$(a(x,D)u,v) = \int\int\int e^{\mathrm{i}(x-y)\cdot\xi}a(x,\xi)u(y)\overline{v(x)}\mathrm{d}x\mathrm{d}y\frac{\mathrm{d}\xi}{(2\pi)^n}.$$

Integrating by parts with respect to ξ, show that

$$(a(x,D)u,v) = \int\int e^{\mathrm{i}x\xi}b(x,\xi)\tilde{u}(x,\xi)\overline{v(x)}\mathrm{d}x\mathrm{d}\xi,$$

where $b \in C_0^\infty(\mathbb{R}^n\times\mathbb{R}^n)$ satisfies $\sum_{a\in\{0,1\}^n} \sup|\partial_x^\alpha b| \leq C\|a\|$.

d) Writing $v(x) = \int e^{\mathrm{i}x\cdot\eta}\hat{v}(\eta)\frac{\mathrm{d}\eta}{(2\pi)^n}$, in the above integral, and integrating by parts with respect to x, show that

$$(a(x,D)u,v) = \sum_{\alpha\in\{0,1\}^n} \int\int e^{\mathrm{i}x\xi} c_\alpha(x,\xi)\partial_x^\alpha \tilde{u}(x,\xi)\overline{\tilde{v}(x,\xi)}\mathrm{d}x\mathrm{d}\xi,$$

where $c_\alpha \in C_0^\infty(\mathbb{R}^n \times \mathbb{R}^n)$, $\sup|c_\alpha| \le C\|a\|$. Conclude.

Remark. This exercise of course proves a lot more than Theorem 5.1: it gives the L^2 continuity of operators of the class $S^0_{0,0}$ (see Exercise 2.9 and [Hw]) to which the proof of Theorem 5.1 cannot be adapted. On the other hand, Exercise 4.12 exhibits a symbolic calculus for the class $S^m_{\rho,\delta}$ when $1 \ge \rho > \delta \ge 0$, which enables us to obtain the L^2 continuity of the symbol operators in $S^0_{\rho,\delta}$.

When $1 > \rho = \delta > 0$, the proof in Exercise 5.3 can be adapted [Hw] while L^2 continuity is not guaranteed in the other cases $0 \le \rho < \delta \le 1$ and $\rho = \delta = 1$ (see Coifman and Meyer [CM]).

Exercise 5.4 (*)**.** (Singular integrals) Let $k \in C^0(\mathbb{R}^n \setminus \{0\})$ be locally Lipschitz and such that

$$\forall x \in \mathbb{R}^n, \quad |k(x)| \le \frac{C}{|x|^n}, \quad |\partial_x k(x)| \le \frac{C}{|x|n+1}.$$

($k(x-y)$ thus satisfies some of the estimates established for the kernel of a pseudo-differential operator of order 0, away from the diagonal $\{x=y\}$ (see Exercise 3.2).)

It is proposed to study the existence and L^2 continuity of the 'principal value' operator

$$Au(x) = \lim_{\varepsilon\to 0+} \int_{|x-y|\ge\varepsilon} k(x-y)u(y)\mathrm{d}y.$$

a) Show that Au exists in $\mathcal{D}'(\mathbb{R}^n)$ for all $u \in C_0^\infty(\mathbb{R}^n)$ if and only if the following condition is realized:

$$(*) \qquad \int_{\varepsilon\le|x|\le 1} k(x)\mathrm{d}x \text{ has a limit when } \varepsilon \text{ tends to } 0.$$

(Compare $Au(x)$ with $u(x)\int_{\varepsilon\le|y|\le 1} k(y)\mathrm{d}y$.) Verify that the family of functions $k_\varepsilon = k\cdot 1_{\{|x|\ge\varepsilon\}}$ then converges in $\mathcal{S}'(\mathbb{R}^n)$ to a distribution denoted by k_0 and that A is bounded on L^2 if and only if $\hat{k}_0 \in L^\infty(\mathbb{R}^n)$.

b) Verify that $\hat{k}_\varepsilon \in L^2(\mathbb{R}^n)$, and that $|\hat{k}_\varepsilon(\xi)| \le C$ for all $\varepsilon > 0$, for all ξ such that $|\xi| \ge 1/\varepsilon$. (One might estimate $\xi_j\hat{k}_\varepsilon(\xi)$, for $j = 1,\dots,n$.)

c) If $|\xi| \le 1/\varepsilon$, set $\hat{k}_\varepsilon(\xi) = I_\varepsilon(\xi) + J(\xi)$, where

$$J(\xi) = \int_{|x|\ge 1/|\xi|} k(x)e^{-\mathrm{i}x\cdot\xi}\mathrm{d}x.$$

Verify that $|J(\xi)| \leq C$ and that

$$\left| I_\varepsilon(\xi) - \int_{\varepsilon \leq |x| \leq 1/|\xi|} k(x)\mathrm{d}x \right| \leq C.$$

Deduce that, under the condition $(*)$, the operator A is bounded on L^2 if and only if the following condition is realized:

$(**)$ $$\int_{1 \leq |x| \leq R} k(x)\mathrm{d}x \text{ remains bounded as } R \to +\infty.$$

What happens to $(*)$ and $(**)$ when k is homogeneous of degree $-n$?

d) Verify that the operator A associated with $k(x) = \frac{|x|^{\mathrm{i}\gamma - n}}{1 + |\log |x||}$ is well defined and bounded on L^2. ($\gamma \in \mathbb{R} \setminus \{0\}$.)

e) Take $k(x) = |x|^{\mathrm{i}\gamma - n}$. Verify that k does not satisfy the condition $(*)$, but that for any complex number α of modulus 1, one can define

$$A_\alpha u(x) = \lim_{n \to \infty} \int_{|x-y| \geq \varepsilon_n} k(x-y)u(y)\mathrm{d}y,$$

where $\varepsilon_n \to 0^+$ and $\varepsilon_n^{\mathrm{i}\gamma} \to \alpha$. Verify that $A_\alpha - A_\beta = \frac{\alpha - \beta}{\mathrm{i}\gamma}$ and that the A_α are bounded on L^2.

Exercise 5.5. Let $B(L^2)$ be the vector space of bounded operators on L^2, with the natural Banach space structure.

a) Verify that the mapping

$$\begin{aligned} S^0 &\to B(L^2), \\ a &\mapsto a(x, D) \end{aligned}$$

is continuous. Is its image closed?

b) Let $a \in S^0$. Prove that the mapping

$$\begin{aligned} [0,1] &\to B(L^2), \\ \varepsilon &\mapsto a(x, \varepsilon D) \end{aligned}$$

is not in general continuous, but that it is continuous if $B(L^2)$ has the simple convergence topology.

Exercise 5.6. If s is an integer, prove that

$$H^s = \{u \in L^2, D^\alpha u \in L^2, |\alpha| \leq s\}.$$

Exercise 5.7. If $s > n/2$, prove that H^s is continuous in the space of continuous functions tending to 0 at infinity and that the following holds:

$$\sup_{x \in \mathbb{R}^n} |u(x)| \leq C|u|_s.$$

Exercise 5.8. To which Sobolev space does δ belong in $\mathbb{R}^n$?

Exercise 5.9. This exercise determines all P-local pseudo differential operators, i.e. such that $u \subset \operatorname{supp} u$ for all $u \in \mathcal{S}$.

For example, differential operators are local. We shall show that there are no others. (This is also a consequence of a general theorem due to Peetre.)

a) Let P be a local pseudo-differential operator of order $m < -n/2$. Let $x_0 \in \mathbb{R}^n$, and let $u \in \mathcal{S}$. Show that there exists a sequence $u_k \in \mathcal{S}$ such that $u_k = 0$ near x_0 for all k, and $u_k \to u$ in $L^2(\mathbb{R}^n)$. Deduce that $Pu(x_0) = 0$. (Use Exercise 5.7.)

b) Suppose P is local of order < 0. Show that $P = 0$. (Use Exercise 4.2.)

c) Suppose P is local of order $m < k \in \mathbb{N}$. Suppose that $f_1, \ldots, f_k$ in C^∞ on $\mathbb{R}^n$ are bounded together with their derivatives. Show that $\operatorname{Ad} f_1 \ldots \operatorname{Ad} f_k P = 0$ (recall that $(\operatorname{Ad} A)B = [A, B] = AB - BA$). Deduce that if $u \in \mathcal{S}$ satisfies $u^{(j)}(x_0) = 0$ for $j \leq k-1$, then $Pu(x_0) = 0$. Conclude that $P = \sum_{|\alpha| \leq k-1} a_\alpha D^\alpha$ where the a_α are C^∞ and bounded together with their derivatives. (One might use Taylor's formula.)

Exercise 5.10. Let $r \in S^{-\infty}$, such that $\forall x \in \mathbb{R}^n$, $r(x, 0) = 1$. Set, for $\varepsilon > 0$, $\rho_\varepsilon = r(x, \varepsilon D)$.

a) Prove the inequalities $|\rho_\varepsilon u|_0 \leq C|u|_0$, $|u - \rho_\varepsilon u|_0 \leq C_m \varepsilon^m |u|_m$ for all $u \in \mathcal{S}$, $m > 0$. Deduce that for all $u \in L^2$, $\rho_\varepsilon u \to u$ in L^2 when $\varepsilon \to 0$ and that, if A has negative order, $\rho_\varepsilon A$ tends to A in norm.

b) Let L be a pseudo-differential operator of order 1. Prove that $|[L, \rho_\varepsilon]u|_0 \leq C|u|_0$ for all $u \in \mathcal{S}$. Denote $D(L) = \{u \in L^2, Lu \in L^2\}$ and suppose $D(L)$ has the norm $|u|_0 + |Lu|_0$. Show that $D(L)$ is a Hilbert space in which $H^{+\infty}$ and then $\mathcal{S}$ are dense.

Exercise 5.11. Let P be a pseudo-differential operator of order 1. For $s \in \mathbb{R}$ and $k \in \mathbb{N}$, set

$$H^{s,k}(P) = \{u \in H^2, P^j u \in H^2 \text{ for all } j \leq k\}.$$

a) Prove that if A is a pseudo-differential operator of order 0, then $u \in H^{s,k}(P)$ implies $Au \in H^{s,k}(P)$. (In particular, $u \in H^s$ and $Pu \in H^{s+k-1}$ implies $Au \in H^{s,k}(P)$.)

b) Let $p, q \in S^1$ be such that there exists $a \in S^0$ satisfying $p - aq \in S^0$. Show that $H^{s,k}(q(x, D)) \subset H^{s,k}(p(x, D))$.

Exercise 5.12 (Elliptic operators and the Gårding inequality). Let $a \in S^0$.

a) Suppose that $a(x, D)$ satisfies a Gårding inequality

$$\forall u \in \mathcal{S},\ \operatorname{Re}(a(x, D)u, u) + C|u|^2_{-\delta} \geq c_0 |u|^2_0.$$

Show that $\operatorname{Re} a(x, \xi) \geq c_0$ for $|\xi|$ sufficiently large (use Exercise 5.1).

b) Suppose that $a(x, D)$ satisfies an elliptic type inequality

$$|u|_0 \leq C(|a(x, D)u|_0 + |u|_{-\delta}).$$

Prove that a is elliptic of order 0.

c) Revisit a) and b) with $a \in S^m$, $m \in \mathbb{R}$.

Exercise 5.13. Let $a \in S^m$ be a symbol satisfying $\operatorname{Re} a(x, \xi) \geq C|\xi|^m$ for ξ sufficiently large and k an integer > 1. Show that there exists $b \in S^{m/k}$ such that $b(x, D)^k = a(x, D)$ modulo an operator of order $-\infty$.

Exercise 5.14. Let $a \in S^0$ be such that $|a(x, \xi)| \geq C > 0$ for all $(x, \xi) \in \mathbb{R}^n \times \mathbb{R}^n$. We shall show that, for $\varepsilon > 0$ sufficiently small, $a(x, \varepsilon D)$ is invertible. (Note that this is not a consequence of the fact that $|a(x, 0)| \geq C$; see Exercise 5.5, part b).)

a) For $\varepsilon \in [0, 1]$, set $a_\varepsilon(x, \xi) = a(x, \varepsilon\xi)$. Show that $q_\varepsilon = \frac{a_\varepsilon - a_0}{\varepsilon}$ is a bounded family of S^1 for $\varepsilon \in]0, 1[$.

b) Verify that $a_\varepsilon \# 1/a_\varepsilon = 1 + \varepsilon(q_\varepsilon \# 1/a_\varepsilon - q_\varepsilon/a_\varepsilon)$ and deduce that there exists $\varepsilon_0 > 0$ such that $\forall \varepsilon \in]0, \varepsilon_0[$, $a_\varepsilon(x, D)$ is an isomorphism of L^2.

Exercise 6.1 (Partitions of unity). Let Ω be an open set in $\mathbb{R}^n$ and $(U_i)_{i \in I}$ a family of relatively compact open sets covering Ω. We shall show that there exists a family of functions $(\varphi_i)_{i \in I}$ such that

i) $\forall i$, $\varphi_i \in C_0^\infty(U_i)$, $0 \leq \varphi_i \leq 1$.

ii) The family $(\operatorname{supp} \varphi_i)$ is locally finite. (That is to say that, for any compact subset K of Ω, $K \cap \operatorname{supp} \varphi_i = \emptyset$ except for a finite number of $i \in I$.)

iii) In Ω, $\sum_{i \in I} \varphi_i = 1$. (This sum is well defined, following ii).)

a) Prove that we may assume that $I = \mathbb{N}$.

b) Suppose that the family (U_i) is locally finite. Show that there exists a family of open sets (V_i) such that $\forall i \in I$, $\overline{V}_i \subset U_i$ and $\Omega = \bigcup_{i \in I} V_i$. (One might construct the V_i recursively so that $\forall n \in \mathbb{N}$, $\Omega = \bigcup_{i \leq n} V_i \cup \bigcup_{j > n} U_j$.) Then choose $\psi_i \in C_0^\infty(U_i)$, $\psi_i \geq 0$, such that $\psi_i > 0$ on V_i, and verify that $\varphi_i = \psi_i / \sum_{j \in I} \psi_j$ satisfies the question.

c) General case. Construct a locally finite sequence (A_n) of relatively compact open subsets of Ω, covering Ω. For each n, choose a finite $I_n \subset I$ such that $\overline{A}_n \subset \bigcup_{i \in I_n} U_i$, and set $U'_{n,i} = U_i \cap A_n$ if $i \in I_n$.

Show that the $U'_{n,i}$, $n \in \mathbb{N}$, $i \in I_n$, form a finite local covering of Ω: choose a partition of unity $(\varphi'_{n,i})$ as in b), and set $\varphi_j = \sum_{n / i \in I_n} \varphi'_{n,i}$.

In the exercises which follow, we let $\psi^m_{ps}(\Omega)$ denote the space of properly supported pseudo-differentials on Ω, of order m.

Exercise 6.2. a) Show that Proposition 2.3 generalizes to the case of elements of $S^{+\infty}_{\mathrm{loc}}(\Omega \times \mathbb{R}^n)$

b) Generalize Theorems 3.2.3 and 4.1 of the symbolic calculus to the class of properly supported operators on Ω.

c) Let $a \in S^m_{\mathrm{loc}}(\Omega \times \mathbb{R}^n)$, and let $A \in \Psi^m_{ps}(\Omega)$ have symbol a. Show that there exists $B \in \Psi^m_{ps}(\Omega)$ satisfying $BA = I + R$ (resp. $AB = I + R$), where R is a C^∞ kernel, if and only if a satisfies the following condition: for any compact subset K of Ω, there exist $C_K > 0$ and R_K such that $\forall x \in K$, $|a(x,\xi)| \geq C_K|\xi|^m$ if $|\xi| \geq R_K$. (We then say that a (or A) is elliptic of order m.)

d) Suppose $p_m(x,\xi)$, $(x,\xi) \in \Omega \times \mathbb{R}^n \setminus \{0\}$ is a homogeneous function in ξ of degree m. Suppose $P \in \Psi^m_{ps}(\Omega)$ with symbol p is such that $p - \tilde{p}_m \in S^{m-1}_{\mathrm{loc}}$ ($\tilde{p}_m \in C^\infty(\Omega \times \mathbb{R}^n)$ is a function which coincides with p_m for $|\xi| \geq 1$): p_m is said to be the principal symbol of P. (See Section 7.2.2.) Under what conditions on p_m is the operator P elliptic of order m?

Exercise 6.3. a) Let $A \in \Psi^m_{ps}(\Omega)$. Show that, for any compact subset K of Ω, for any real s, there exists $C_{k,s} > 0$ such that

$$\forall u \in C_0^\infty(K),\ |Au|_s \leq C_K|u|_{s+m}.$$

b) Denote

$$H^s_{\mathrm{loc}}(\Omega) = \{u \in \mathcal{D}'(\Omega), \forall \varphi \in C_0^\infty(\Omega), \varphi u \in H^s(\mathbb{R}^n)\},$$

which has the semi-norms $|\varphi u|_s$, $\varphi \in C_0^\infty(\Omega)$. Show that the inequality of a) means that A is bounded from $H^{s+m}_{\mathrm{loc}}(\Omega)$ to $H^s_{\mathrm{loc}}(\Omega)$.

Exercise 6.4. This exercise provides a characterization of local pseudo-differential operators via their action on L^2 (see [CM]).

a) Let T be a bounded linear operator on $L^2(\mathbb{R}^n)$, with kernel of compact support in $\mathbb{R}^n \times \mathbb{R}^n$. For $\xi \in \mathbb{R}^n$, let e_ξ denote the function defined by $e_\xi(x) = e^{\mathrm{i}x\cdot\xi}$, and set

$$\sigma_T(x,\xi) = T(e_\xi)(x)e^{-\mathrm{i}x\cdot\xi}.$$

Justify the existence of σ_T, and verify that it is continuous in ξ, with values in $L^2(\mathbb{R}^n_x)$. Finally, verify that

$$\sup_{\xi\in\mathbb{R}^n} \int |\sigma_T(x,\xi)|^2 \mathrm{d}x < +\infty$$

and that, $\forall u \in C_0^\infty(\mathbb{R}^n)$, $Tu(x) = \int e^{\mathrm{i}x\cdot\xi}\sigma_T(x,\xi)\hat{u}(\xi)\frac{d\xi}{(2\pi)^n}$ (note, for example, that for any function $\varphi \in C_0^\infty(\mathbb{R}^n)$,

$$\langle Tu, \varphi\rangle = \langle u, T^*\varphi\rangle = (2\pi)^{-n}\int \hat{u}(\xi)\langle e_\xi, T^*\varphi\rangle \mathrm{d}\xi).$$

b) Let $\mathcal{T}$ denote the vector space of operators T, as in a), such that, for any finite family of vector fields $X_1, \dots, X_k$, the operator $ad(X_1)\dots ad(X_k)T$ is bounded on $L^2(\mathbb{R}^n)$.

i) Show that any pseudo-differential operator of order 0 with kernel supported in a compact set is in $\mathcal{T}$.

ii) Show that if T_1, T_2 are in $\mathcal{T}$, then T_1T_2 is in $\mathcal{T}$.

iii) Denote $\Sigma = \{\sigma_T, T \in \mathcal{T}\}$. Show that Σ is stable under multiplication by a $C^\infty(\mathbb{R}^n)$ function, and under the action of $\partial/\partial x_j$, $\partial/\partial\xi_j, \xi_k\partial/\partial\xi_j$, for all j, k in $\{1, \dots, n\}$.

iv) Using a), show that $\Sigma \subset S^0(\mathbb{R}^n \times \mathbb{R}^n)$.

c) State and prove a result analogous to b) in the framework of properly supported operators on a compact subset Ω of $\mathbb{R}^n$.

d) Let $\sigma(\xi) = \exp(ie^{\langle\xi\rangle})e^{-\langle\xi\rangle^{1/2}}$, $\langle\xi\rangle = (1 + |\xi|^2)^{1/2}$. Show that the operator $\sigma(D)$ is bounded from H^s to H^t for all reals s and t and that for any finite family of vector fields $X_1, \dots, X_k$ with symbols in S^1, the operator $ad(X_1)\dots ad(X_k)\sigma(D)$ is bounded on $L^2(\mathbb{R}^n)$. Does the symbol σ belong to S^0?

Exercise 6.5 $(*)$. a) Let s, t be two elements of $[0, 1[$ satisfying $s + t < 1$. Let $A \in \Psi^{-m}_{ps}(\Omega)$, $m = s + t$, P and $Q \in \Psi^1_{ps}(\Omega)$: suppose that P and Q both have a real principal symbol.

Show that, for any compact subset K of Ω, there exists $C_K > 0$ such that

$$\forall u \in C_0^\infty(K), \quad |(Au, PQu)| \le C_K(|u|_0^2 + |Pu|_{-s}^2 + |Qu|_t^2).$$

Choosing $A = E^*E[P, Q]$, where $E \in \Psi^{-(m+1)/2}$ is elliptic, deduce that

$$\forall u \in C_0^\infty(K), \quad |[P, Q]u|_{-(1+s+t)/2} \le C_K(|u|_0 + |Pu|_{-s} + |Qu|_{-t})$$

(with different constants C_K).

b) Let $X_1, \dots, X_p$ be real vector fields which are C^∞ on Ω. Suppose that at each point x of Ω, the vectors $X_i(x)$, $[X_j, X_k](x)$ for $1 \le i \le p$, $1 \le j < k \le p$, generate $\mathbb{R}^n$.

i) Using a) with $s = t = 0$, derive the estimate

$$\forall u \in C_0^\infty(K), \quad |u|_{1/2} \le C_K\left(|u|_0 + \sum_{i=1}^p |X_iu|_0\right).$$

ii) Now let $u \in L^2_{\text{loc}}(\Omega)$ be such that, for all i, $X_iu \in L^2_{\text{loc}}(\Omega)$. Applying the above inequality to $R_\varepsilon u$, with

$$R_\varepsilon = \varphi\chi(\varepsilon D)\varphi, \varphi \in C_0^\infty(\Omega), \chi \in S \text{ and } \chi(0) = 1,$$

conclude that $u \in H^{1/2}_{\text{loc}}(\Omega)$. (One may remark that $R_\varepsilon = r_\varepsilon(x, D)$, where the family (r_ε) is bounded in $S^0(\mathbb{R}^n \times \mathbb{R}^n)$.)

iii) Let $u \in \mathcal{D}'$, such that for all i, $X_i u \in H^s_{\text{loc}}(\Omega)$. Show that $u \in H^{2+1/2}_{\text{loc}}(\Omega)$. (One may remark that each $x \in \Omega$ has an open neighbourhood ω such that $u \in H^\sigma_{\text{loc}}(\omega)$ for some σ.) Show that if $X_i u \in C^\infty(\Omega)$, $1 \le i \le p$, then $u \in C^\infty(\Omega)$.

iv) In $\mathbb{R}^2$, consider $X_1 = \partial/\partial x$, $X_2 = x\partial/\partial y$. Let φ, ψ be two C^∞ functions with compact support on $\mathbb{R}$, with $\operatorname{supp}\psi \subset [1,2]$. For $\lambda \ge 1$, set $u_\lambda(x,y) = \psi(\lambda x)\int_0^{+\infty} e^{iy\eta}\varphi(x\sqrt{\eta})\mathrm{d}\eta$. Estimate

$$\int_{-\infty}^{+\infty} \|u_\lambda(x,\cdot)\|^2_H \mathrm{d}x$$

for $s \ge 0$, when $\lambda \to +\infty$, then $|X_i u_\lambda|_0^2$, and conclude that the above result is optimal.

c) Again take $X_i, \dots, X_p$ on Ω, but this time suppose that for each $x \in \Omega$, $\mathbb{R}^n$ is generated by the $X_I(x)$, $|I| \le r$, where

$$\text{if } I = (i_1, \dots, i_q) \in \{1, \dots, p\}^q,\ q = |I|,$$
$$\text{then } X_I = (adX_{i_1})\dots(adX_{i_{q-1}})X_{i_q}.$$

Arguing as in b), show that there exists $\delta > 0$ such that, if $u \in \mathcal{D}'(\Omega)$ satisfies $X_i u \in H^s_{\text{loc}}(\Omega)$, $1 \le i \le p$, then $u \in H^{s+\delta}_{\text{loc}}(\Omega)$.
Remark. Unlike in b), the gain δ exhibited by this method is not optimal. (One can show that the optimal gain is $1/r$.)

Exercise 6.6. Let $X_1, \dots, X_p$ be C^∞ vector fields on Ω. Set

$$P = \sum_{i=1}^p X_i^2.$$

a) Derive the following estimate: for any compact subset K of Ω, all $s \in \mathbb{R}$, all $u \in C_0^\infty(K)$,

$$\sum_{i=1}^p |X_i u|_s^2 \le |\mathrm{Re}\,(Pu, u)_s| + C_{K,s}|u|_s^2,$$

where $(v, w)_s$ denotes the scalar product in $H^s(\mathbb{R}^n)$.

b) Arguing as in Exercise 6.5 b)ii), show that if $u \in H^s_{\text{loc}}(\Omega)$ satisfies $Pu \in H^s_{\text{loc}}(\Omega)$, then $X_i u \in H^s_{\text{loc}}(\Omega)$ for $1 \le i \le p$.

c) Suppose that the X_i satisfy the hypotheses of Exercise 6.5 c). Let $u \in \mathcal{D}'(\Omega)$. What can one say about u if $Pu \in H^s_{\text{loc}}(\Omega)$, if $Pu \in C^\infty(\Omega)$?

Exercise 6.7 $(*)$. Suppose $P \in \Psi^1_{ps}(\Omega)$ has a principal symbol p_1 satisfying

$$\forall x \in \Omega, \forall \xi \in \mathbb{R}^n \setminus \{0\},\ p_1(x,\xi) = 0 \Rightarrow \{\mathrm{Re}\,p_1, \mathrm{Im}\,p_1\}(x,\xi) > 0.$$

a) Let E be a properly supported operator of order -1 with principal symbol $|\xi|^{-1}$. Show that there exists a positive constant C_0 such that the operator

$$Q = [P^*, P] + C_0 E P^* P$$

satisfies the hypotheses of the Gårding inequality.

b) Prove that, for any compact subset K of Ω, there exists $C_K > 0$ such that

$$\forall u \in C_0^\infty(K), \quad |u|_{1/2} \leq C_K(|Pu|_0 + |u|_0).$$

Using a regularizer, as in Exercise 6.5 b), deduce that if $u \in L^2_{\text{loc}}(\Omega)$ and $Pu \in L^2_{\text{loc}}(\Omega)$, then $u \in H^{1/2}_{\text{loc}}(\Omega)$; conclude that if $u \in \mathcal{D}'(\Omega)$ satisfies $Pu \in H^s_{\text{loc}}(\Omega)$, then $u \in H^{s+1/2}_{\text{loc}}(\Omega)$. (Argue as in Exercise 6.5).

P is said to be sub-elliptic with gain of half a derivative.

Exercise 6.8 (∗). (Single variable transmission condition)

a) Preliminary calculation. For $\mu \in \mathbb{R}$ and $\chi \in C^\infty(\mathbb{R})$ satisfying $\chi(\xi) = 1$ for $\xi \geq 1$, and $\chi(\xi) = 0$ for $\xi \leq 1/2$, set

$$u_\mu(x) = \int_{-\infty}^{+\infty} e^{\mathrm{i}x\xi}\chi(\xi)\xi^\mu \mathrm{d}\xi.$$

We know (see Exercise 4.6) that u_μ is a distribution on $\mathbb{R}$ which is C^∞ away from $x = 0$. Show that if $x > 0$, we have
if $\mu \notin \{-1, -2, -3, \ldots\}$,

$$u_\mu(x) = e^{\mathrm{i}\pi/2(\mu+1)}\Gamma(\mu+1)x^{-\mu-1} \text{ modulo } C^\infty([0, +\infty[),$$

$$u_\mu(-x) = e^{-\mathrm{i}\pi/2(\mu-1)}\Gamma(\mu+1)x^{-\mu-1} \text{ modulo } C^\infty([0, +\infty[),$$

and if $n \in \{1, 2, 3 \ldots\}$

$$u_{-n}(x) = \mathrm{i}^{n+1}x^{n-1}\log x \text{ modulo } C^\infty([0, +\infty[),$$

$$u_{-n}(-x) = (-\mathrm{i})^{n+1}x^{n-1}\log x \text{ modulo } C^\infty([0, +\infty[).$$

Here, Γ denotes the Euler function defined in the usual way for $\operatorname{Re} z > 0$, by $\Gamma(z) = \int_0^\infty e^{-t}t^{z-1}\mathrm{d}t$ and extended meromorphically to the whole of the complex plane. *Hint*: First consider the case $\mu > -1$, and compare u_μ with $\int_0^\infty e^{\mathrm{i}x\xi}\xi^\mu \mathrm{d}\xi$. Then remark that $D_x u_\mu = u_{\mu+1}$.

b) Let $\lambda \sim \sum \lambda_j$ be a classical symbol of order m on $\mathbb{R}$. Set

$$u(x) = \int_{-\infty}^{+\infty} e^{\mathrm{i}x\xi}\lambda(\xi)\mathrm{d}\xi.$$

Show that $u|_{x>0}$ extends to a C^∞ function if and only if $\forall j \in \mathbb{N}, \lambda_j(-1) = e^{\mathrm{i}\pi(m-j)}\lambda_j(1)$, which means that $\lambda_j(\xi)$ extends holomorphically to $\mathrm{Im}\xi > 0$.

c) Let $a \sim \sum_{j\geq 0} a_j$ be a classical symbol of order m on $\mathbb{R}$. The operator $a(x, D)$ is said to satisfy the transmission property with respect to $]0, +\infty[$ if $\forall u \in C_0^\infty([0, +\infty[)$, the function $a(x, D)(u^0)|_x > 0$ extends to a C^∞ function. (u^0 denotes the extension of u to all of $\mathbb{R}$ given by $u^0(x) = 0$ if $x < 0$.)

Show that, if $u \in C_0^\infty([0, +\infty[)$,

$$a(x, D)u^0(x) = \int_{-\infty}^{+\infty} e^{\mathrm{i}x\xi}\lambda(\xi)\mathrm{d}\xi,$$

where λ is a classical symbol of order m satisfying, in particular, $\lambda_0(\xi) = -\mathrm{i}\xi^{-1}a_0(0, \xi)u(0)$. (One may begin by remarking that $\hat{u}^0(\xi)$ is a classical symbol of order -1, whose asymptotic expansion can be calculated; then refer to Theorem 2 of the Appendix to calculate λ as a function of a.)

Deduce that $a(x, D)$ satisfies the transmission property whenever the following conditions are satisfied:

$$(*) \qquad \forall k \in \mathbb{N}, \forall j \in \mathbb{N}, \partial_x^k a_j(0, -1) = e^{\mathrm{i}\pi(m-j)}\partial_x^k a_j(0, 1).$$

d) Show, conversely, that if $a(x, D)$ satisfies the transmission property, then the relation $(*)$ holds. (Begin by proving the result for the case $j = k = 0$. Then remark that if $a(x, D)$ satisfies the transmission property, then so does $[D, a(x, D)]$.)

Exercise 7.1 $(*)$. (Stationary phase formula)

a) Let q be a non-degenerate quadratic form on $\mathbb{R}^n$, and let $u \in C_0^\infty(\mathbb{R}^n)$. It is proposed to study the asymptotic behaviour of

$$I(\lambda) = \int e^{\mathrm{i}\lambda q(x)/2}u(x)\mathrm{d}x \text{ as } \lambda \to +\infty.$$

For this, set $x = y/\sqrt{\lambda}$ in the integral, and expand u in the neighbourhood of 0: by estimating the residue using Theorem 1, show that one obtains an asymptotic expansion in powers of λ, whose individual terms are calculated using Exercise 4.9 b):

$$I(\lambda) \sim \left(\frac{2\pi}{\lambda}\right)^{n/2} \frac{e^{\mathrm{i}\sigma\pi/4}}{|\det q|^{1/2}} \sum_{k=0}^{\infty}(2\mathrm{i}\lambda)^{-k}\left(\frac{\hat{q}(D)^k}{k!}u\right)(0),$$

where σ is the signature of q.

Now suppose $f \in C^\infty(\mathbb{R}^n)$ is real-valued, with a unique critical point x_0, where the Hessian $q = f''(x_0)$ is non-degenerate. For $u \in C_0^\infty$, it is proposed to study

$$I_f(\lambda) = \int e^{\mathrm{i}\lambda f(x)}u(x)\mathrm{d}x.$$

b) Morse's Lemma. Show that there exist a neighbourhood V of x_0 and a local diffeomorphism φ of V on a neighbourhood of 0, such that

$$\forall x \in V, \quad f(x) = \frac{1}{2}q(\varphi(x)).$$

For this, look for $\varphi(x)$ in the form $B(x)(x - x_0)$ where $B(x_0) = I$. We remark that we can write

$$f(x) = \frac{1}{2}\langle A(x)(x - x_0), x - x_0\rangle$$

with $A(x)$ symmetric and $A(x_0)$ being the matrix for q, and thus, we reduce to solving

$$B(x)^* A(x_0) B(x) = A(x), \quad B(x_0) = I,$$

for x close to x_0. Conclude by applying the local inversion theorem (for example, in the version stated in Section III.A.1.1) to the mapping $B \mapsto B^* A(x_0) B$ close to $B = I$.

c) Deduce from a) and b) that $I_f(\lambda)$ has an asymptotic expansion in powers of λ with first term

$$\left(\frac{2\pi}{\lambda}\right)^{n/2} e^{i\sigma\pi/4} |\det f''(x_0)|^{-1/2} u(x_0),$$

where σ is the signature of the Hessian $f''(x_0)$. (One might use Lemma 1 to localize the domain of integration around x_0.)

Remark. It is possible to calculate explicitly all the terms of the asymptotic expansion exhibited in c). (See Theorem 7.7.6 of [H1], where a method independent from Morse's lemma is used.)

Throughout what follows, M denotes a C^∞ manifold (countable union of compact sets).

Exercise 7.2. Verify that the partition of unity theorem (Exercise 6.1) remains true on the manifold M.

Exercise 7.3. Choose a positive density μ on M. (If necessary see the note on densities at the end of the exercises.)

a) Let $A \in \Psi^m(M)$ and let $\varphi, \psi \in C_0^\infty$ be such that $\operatorname{supp}\varphi \cap \operatorname{supp}\psi = \emptyset$. Show that there exists $k \in C_0^\infty(M \times M)$ such that

$$\forall u \in C^\infty(M), \forall x \in M, \quad \varphi A \psi u(x) = \int_M k(x, y) u(y) \mu(y).$$

(One might first study the case in which $\operatorname{supp}\varphi \subset U$, $\operatorname{supp}\psi \subset V$, where U and V are disjoint coordinate patches; then $U \cup V$ is a coordinate patch.)

b) Show that $\Psi^{-\infty}(M)$ coincides exactly with the operators of type

$$Ku(x) = \int_M k(x, y) u(y) \mu(y) \quad (u \in C_0^\infty(M)),$$

where $k \in C^\infty(M \times M)$.

Exercise 7.4. a) Let U be an open subset of M. If $A \in \Psi^m(M)$, show that the operator $A|_U$:

$$\begin{array}{rcl} C_0^\infty(u) & \to & C^\infty(U) \\ u & \mapsto & (Au)|_U \end{array}$$

is an element of $\Psi^m(U)$. If $B \in \Psi^m(U)$, $\varphi, \psi \in C_0^\infty(U)$, show that $\varphi B \psi$ defines an element of $\Psi^m(M)$.

b) Let $A : C_0^\infty(M) \to C^\infty(M)$ be a continuous linear operator. Assume that there exists an open covering (U_i) of M such that $A|_{U_i} \in \Psi^m(U_i)$ for all i. Is A an element of $\Psi^m(M)$? (It may suffice to study the case of two open sets in $\mathbb{R}$!) Show that one can conclude, if one also imposes the condition that A satisfies the property exhibited in Exercise 7.3 a).

c) Let $(U_i)_{i \in I}$ be an open covering of M, and for all $i \in I$, let $A_i \in \Psi^m(M)$. We also take $\ell \in [-\infty, m[$, and suppose that for all i, j,

$$A_i|_{U_i \cap U_j} - A_j|_{U_i \cap U_j} \in \Psi^\ell(U_i \cap U_j).$$

Show that there exists $A \in \Psi^m(M)$ such that, for i,

$$A|_{U_i} - A_i \in \Psi^\ell(U_i).$$

Show that A is unique modulo $\Psi^\ell(M)$. In other words, Ψ^m/Ψ^ℓ is a sheaf on M. (Hint: one might take $A = \sum_i \psi_i A_i \varphi_i$, where $\varphi_i, \psi_i \in C_0^\infty(U_i)$ are well chosen. . . .)

d) Apply the result of c) to the following situations:

i) If $a \in S^m(T^*M)$ is m-homogeneous, there exists $A \in \Psi^m(M)$ with principal symbol a.

ii) Let $(m_j)_{j \in \mathbb{N}}$ be a sequence of reals decreasing to $-\infty$ and, for all j, let $A_j \in \Psi^{m_j}(M)$. Show that there exists $A \in \Psi^m(M)$ (unique modulo $\Psi^{-\infty}(M)$) such that

$$A \sim \sum_{j=0}^{\infty}, \text{ in the sense that } \forall N \in \mathbb{N},\ A - \sum_{j<N} A_j \in \Psi^{m_N}(M).$$

iii) Let $A \in \Psi^m(M)$ have principal symbol a_m. Suppose that $a_m \neq 0$ on T^*M. Show that there exists $B \in \Psi^{-\infty}(M)$ such that $AB \equiv I \equiv BA \bmod \Psi^{-\infty}(M)$.

Exercise 7.5. This exercise studies the generalization of Theorem 3.2.3 to manifolds. Let μ be a positive density on M, and let $A \in \Psi^m(M)$. It is proposed to show that there exists $A^* \in \Psi^m(M)$ such that

$$\forall u \in C_0^\infty(M), \forall v \in C_0^\infty(M), \qquad \int_M Au \cdot \overline{v}\mu = \int_M u \cdot \overline{A^* v}\mu.$$

a) Verify that, if such an operator exists, then it is unique among operators $C_0^\infty(M) \to C^\infty(M)$ (or among operators from $C_0^\infty(M)$ to the space of distributions on M, for readers who know what that is).

b) Let $\varphi, \psi \in C_0^\infty(U)$, where U is a coordinate patch. Construct $(\varphi A \psi)^*$. Deduce a construction of $(A\psi)^*$ using a partition of unity subordinate to a covering (U_i) such that $U_0 = U$, $U_i \cap \operatorname{supp} \psi = \emptyset$ for $i \neq 0$. (Use Exercise 7.3 a).) Finally, deduce a construction of A^*.

c) Show that if A has a principal symbol a_m, then A^* has principal symbol $\overline{a}_m$.

Exercise 7.6. a) Demonstrate the invariance under change of coordinates of the fact that an operator is classical in the sense of Section 2. Deduce that one can define classical pseudo-differential operators on M.

b) Let $A \in \Psi^m(M)$ be classical, with principal symbol a_m. Let $x \in M$, $\varphi \in C_0^\infty(M)$, such that $\varphi(x) = 1$, and $\psi \in C^\infty(M)$ such that $\mathrm{d}\psi(x) \neq 0$. Calculate $\lim_{\lambda \to +\infty} \lambda^{-m} e^{-\mathrm{i}\lambda\psi(x)} A(e^{\mathrm{i}\lambda\psi}\varphi)(x)$ as a function of a_m. (Use Theorem 3 and Exercise 7.3 a).)

Exercise 7.7. The object of this exercise is to characterize the pseudo-differential operators on $M = \mathbb{T}^n = (\mathbb{R}/2\pi\mathbb{Z})^n$. For $k \in \mathbb{Z}^n$, we let e_k denote the C^∞ function on $\mathbb{T}^n$ defined by $e_k(x) = e^{\mathrm{i}k\cdot x}$. If $u \in C^\infty(\mathbb{T}^n)$, we let $\hat{u}(k) = (2\pi)^{-n} \int_{\mathbb{T}^n} u(x) e^{-\mathrm{i}k\cdot x} \mathrm{d}x$. If $a = (a(k))_{k\in\mathbb{Z}^n}$ is a sequence of complex numbers we define the sequence $\delta_j a$ $(1 \leq j \leq n)$ by

$$\delta_j a(k) = a(k + \varepsilon_j) - a(k),$$

where $(\varepsilon_1, \ldots, \varepsilon_n)$ is the canonical basis of $\mathbb{Z}^n$. We then define δ^α for $\alpha \in \mathbb{N}^n$ in the usual way.

a) Let $A \in \Psi^m(\mathbb{T}^n)$. For $x \in \mathbb{T}^n$ and $k \in \mathbb{Z}^n$, we set

$$a(x, k) = e_{-k}(x) A e_k(x).$$

Verify that $|a(x,k)| \leq C(1 + |k|)^m$ (use Theorem 3) and that

$$\begin{aligned} \partial_j a(x,k) &= e_{-k}(x)[\partial_j, A] e_k(x), \\ \delta_j a(x,k) &= e_{-k}(x)(e_{-\varepsilon_j} A e_{\varepsilon_j} - A)(e_k)(x). \end{aligned}$$

Deduce that

$$(*) \qquad \forall \alpha, \forall \beta, \quad |\partial_z^\alpha \delta^\beta a(x,k)| \leq C_{\alpha\beta}(1 + |k|)^{m-|\beta|}.$$

b) Conversely, suppose $a = a(x,k)$ is of class C^∞ in x and satisfies the condition $(*)$. Then we define $A : C^\infty(\mathbb{T}^n) \to C^\infty(\mathbb{T}^n)$ by

$$(**) \qquad Au(x) = \sum_{k\in\mathbb{Z}^n} a(x,k) \hat{u}(k) e^{ik\cdot x}.$$

i) For $\varphi, \psi \in C^\infty(\mathbb{T}^n)$ such that $\operatorname{supp}\varphi \cap \operatorname{supp}\psi = \emptyset$, show that one can write

$$\varphi A\psi u(x) = \int_{\mathbb{T}^n} k(x,y)u(y)\mathrm{d}y, \text{ with } k \in C^\infty(\mathbb{T}^n \times \mathbb{T}^n).$$

ii) We denote $U = (\mathbb{T}^1 \setminus \{1\})^n$ which we identify with $]0, 2\pi[^n$ by the usual map. For $\varphi, \psi \in C_0^\infty(U)$, verify that $\varphi A\psi$ belongs to $\Psi^m(U)$. (Note that $\delta^\alpha(e^{\mathrm{i}k\cdot z}) = z^\alpha e^{\mathrm{i}k\cdot z} q_\alpha(z)$ where q_α is C^∞, non-zero on U.)

iii) Conclude.

The next three exercises show how one can apply pseudo-differential operators to analysis on compact manifolds. In order to better show how pseudo-differential operators work, we have chosen to give self-contained proofs of the classical results on finiteness, duality and spectral decomposition, and the only non-trivial result from functional analysis used here is the Hilbert projection theorem (Exercise 7.9 e)).

Readers who wish to learn about the general concepts of functional analysis which are attached to these results (Fredholm operators, etc.) may refer, for example to Section 19.1 of Hörmander [H5].

In all that follows, we take a compact manifold M and a positive density μ on M. The space $L^2(M)$ has the Hilbert scalar product

$$(u, v) = \int_M u\bar{v}\mu.$$

As usual, we denote $|u|_0^2 = (u, u)$. For the norm thus defined, $C^\infty(M)$ is a dense subspace of $L^2(M)$.

Exercise 7.8 $(*)$. (Finiteness and duality) Let M be a compact manifold of dimension n, and let $P \in \Psi^m(M)$ be an elliptic operator. We choose a positive density μ on M and introduce the prehilbertian scalar product

$$(u, v) = \int_M u\bar{v}\mu$$

on $C^\infty(M)$. In Exercise 7.5, we showed that P had an adjoint P^* for this structure and that $P^* \in \Psi^m(M)$.

Moreover, any function $k \in C^\infty(M \times M)$ which can be written in the form $k(x, y) = \sum_{j=1}^N \varphi_j(x)\psi_j(y)$ where $\varphi_j, \psi_j \in C^\infty(M)$ will be called a kernel of finite rank, where this nomenclature extends to the operator K associated with the latter by the formula $Ku(x) = \int_M k(x, y)u(y)\mu(y)$.

a) Show that if K is a kernel of finite rank, then $\operatorname{Ker}(I - K)$ is finite dimensional and $\operatorname{Im}(I - K) = (\operatorname{Ker}(I - K^*))^\perp$.

b) Let $k \in C^\infty(M \times M)$. Show that, for all $\varepsilon > 0$, there exists k' of finite rank such that $\sup_{M\times M} |k - k'| \leq \varepsilon$. (One might consider the Stone–Weierstrass theorem.)

Deduce that there exists $Q \in \Psi^{-m}(M)$ such that $QP = I - K$, where K is a kernel of finite rank.

c) Show that $\operatorname{Ker} P$ is finite dimensional. Let $f \in (\operatorname{Ker} P^*)^\perp$. Show that $Qf \in (\operatorname{Ker}(I - K^*))^\perp$, and deduce that there exists $g \in C^\infty(M)$ such that $f - Pg \in \operatorname{Ker} Q$. Show that one can choose g such that $\|f - Pg\|$ is minimal and that then $h = f - Pg$ is orthogonal to $P(\operatorname{Ker}(QP))$. Deduce that $P^*h \in \operatorname{Im}(P^*Q^*)$ and then that $h = 0$. Conclude that $\operatorname{Im} P = (\operatorname{Ker} P^*)^\perp$.

Exercise 7.9 (*). Let $A \in \Psi^m(M)$ with $m < 0$. By Theorem 5.1,

$$A : C^\infty(M) \to C^\infty(M)$$

extends to a bounded operator of $L^2(M)$, which we shall again denote by A. Suppose that A is self-adjoint, i.e. $A^* = A$.

a) Let $F \subset L^2(M)$ be a closed subspace which is stable under A and on which A is not identically zero.

i) Let $s(F) = \sup_{u\in F, |u|_0 = 1} |(Au, u)|$. Show that $s(F) \in]0, +\infty[$.

ii) Let (u_j) be a sequence of F such that $|u_j|_0 = 1$ and $(Au_j, u_j) \to \lambda$ with $|\lambda| = s(F)$. Denote $\varepsilon = \lambda/s(F) \in \{-1, 1\}$.

Show that $|Au_j - \lambda u_j|_0 \to 0$ (apply the Schwarz inequality to $Q(v) = ((s(F) - \varepsilon A)v, v))$ and, using Exercise 7.8 b), show that a subsequence of (u_j) converges in norm to a non-zero element of $\operatorname{Ker}(A - \lambda) \cap F$. Verify also that $\operatorname{Ker}(A - \lambda)$ is contained in $C^\infty(M)$ and is finite dimensional.

b) Show that there exist a sequence (E_k) of pairwise orthogonal subspaces of $L^2(M)$, with finite non-zero dimension, and a sequence of real numbers (λ_k), such that

$$\forall k, A_{|E_k} = \lambda_k \mathrm{id} \text{ and } |\lambda_{k+1}| = s(F_k), \text{ with } F_k = \left(\bigoplus_{j=0}^{k} E_j\right)^\perp ;$$

and finally $E_k \subset C^\infty(M)$ if $\lambda_k \neq 0$.

c) Show that $\lim_{k\to\infty} \lambda_k = 0$. (Use Exercises 5.10 a) and 7.8 b) to show that A is the limit in the norm of kernels of finite rank; deduce that A cannot satisfy an inequality of type $|(Au, u)| \geq c|u|^2$ with $c > 0$ on an infinite dimensional subspace.)

d) Show that if (λ_k) is stationary with value 0, then A is a kernel of finite rank.

e) Set $E = \bigoplus_{k=0}^{\infty} E_k + \operatorname{Ker} A$. Show that E is dense in $L^2(M)$. (Study $E^{\perp}$.)

Exercise 7.10 (*). Suppose $P \in \Psi^m(M)$ of order m is formally self-adjoint, i.e. $P^* = P$, and has principal symbol p_m satisfying

$$\forall (x,\xi) \in T^*M, p_m\,(x,\xi) \geq c|\xi|^m, \text{ with } c > 0.$$

a) Show that there exists $a > 0$ such that $P + a$ is injective on $C^\infty(M)$. (Use the Gårding inequality.)

Use Exercise 7.8 to show that $P + a : C^\infty(M) \to C^\infty(M)$ is invertible.

b) Apply the results of Exercise 7.9 to $A = (P+a)^{-1}$ to show that there exist a sequence (E_k) of pairwise orthogonal subspaces of $C^\infty(M)$, of finite non-zero dimensions, and a sequence of reals (λ_k) tending to $+\infty$ such that

$$L^2(M) = \overline{\bigoplus_k E_k} \text{ and } \forall k, P_{|E_k} = \lambda_k \mathrm{Id}.$$

Note. Densities on a manifold. For readers unfamiliar with the calculus on manifolds, we recall the notion of density here in a few lines.

If E is a vector space on $\mathbb{R}$ of dimension n, we let $|\Omega|(E)$ denote the set of mappings $\varphi : E^n \to \mathbb{C}$ such that, for any linear operator A on E, for all vectors $h_1, \dots, h_n$ of E,

$$(*) \qquad \varphi(Ah_1, \dots, Ah_n) = |\det A| \varphi(h_1, \dots, h_n).$$

For example, if $(e_1, \dots, e_n)$ is a basis of E, $\varphi = |e_1^* \wedge \dots \wedge e_n^*|$ is an element of $|\Omega|(E)$ with values in $\mathbb{R}^+$. It is clear that $|\Omega|(E)$ is a vector space over $\mathbb{C}$ of dimension 1 and that such an element constitutes a basis,

When M is a manifold of dimension n, we let $|\Omega|(TM)$ denote the disjoint union of the $|\Omega|(T_mM)$ for $m \in M$: $|\Omega|(TM)$ is organized as a complex vector bundle of rank 1 on M, which we call the bundle of volume elements on M. The term density refers to any section (C^∞) of this bundle; the interest of a density ρ is that it can be integrated on M (at least if it decreases sufficiently at infinity, and in any case if it has compact support). In fact, in a system of local coordinates, one can write

$$\rho = f(x_1, \dots, x_n)|\mathrm{d}x_1 \wedge \dots \wedge \mathrm{d}x_n|, \text{ with } f \in C^\infty,$$

and, if ρ is of compact support in the coordinate patch, one can define

$$\int_M \rho = \int_{\mathbb{R}^n} f(x_1, \dots, x_n) \mathrm{d}x_1 \dots \mathrm{d}x_n,$$

where the formula (*) ensures the independence from the choice of coordinate system thanks to the formula for changes of variables in integrals. By partition of unity $\int_M \rho$ is defined in this way for any density ρ with compact support on M.

Of course, if M is orientable, any density ρ can be written as $f|\omega|$, where $f \in C^\infty(M)$, and ω is an n-form which is everywhere non-zero.

If ω is positive, the quantity $\int_M \rho$ is just $\int_{\vec{M}} f\omega$, where $\vec{M}$ denotes the oriented manifold M.

Densities allow us to integrate on a manifold without this orientability hypothesis. We note that, thanks to partitions of unity, one can always trivialize the bundle $|\Omega|(TM)$, that is, construct a density ρ_0, which is everywhere non-zero: any other density will be written as $f\rho_0$ with $f \in C^\infty(M)$. Generally it is the convention to choose ρ_0 positive, that is to say such that $\rho_0(m)(h_1, \dots, h_n) > 0$ for all $m \in M$, for all independent $h_1, \dots, h_n \in T_mM$. (This choice is easy, taking into account $(*)$.)

Chapter II

Nonlinear dyadic analysis, microlocal analysis, energy estimates

At first sight it might seem a little ridiculous to present 'applications' of a notion such as that of pseudo-differential operators which is encountered very commonly in nature: nevertheless, without wishing to refer the reader to specialist articles or papers, we describe certain aspects of the applications in an elementary and very limited way here. While Sections A and B can be viewed as natural developments (rich in consequences) of the symbolic calculus of the classical pseudo-differential presented in Chapter I, Section C uses the efficiency of the pseudo-differential *tool* to obtain vital hyperbolic energy estimates.

These estimates can be used to obtain Hörmander's theorem on the propagation of singularities (one of the key results of microlocal analysis); in addition, they are central to the nonlinear perturbation techniques presented in Chapter III.

A. Nonlinear dyadic analysis

1. Littlewood–Paley decomposition: general properties.

1.1. *Littlewood–Paley decomposition.* Suppose $\psi \in C_0^\infty(\mathbb{R}^n)$, $\psi(\xi) = 1$ for $|\xi| \leq 1/2$, $\psi(\xi) = 0$ for $|\psi| \geq 1$. Set $\varphi(\xi) = \psi(\xi/2) - \psi(\xi)$: φ is supported in the shell $1/2 \leq |\xi| \leq 2$, and, for all ξ,

$$1 = \psi(\xi) + \sum_{p \geq 0} \varphi(2^{-p}\xi)$$

(there are never more than two non-zero terms in this series). For $u \in \mathcal{S}'(\mathbb{R}^n)$, set

$$u_{-1} = S_0 u = \psi(D)u, \quad u_p = \varphi(2^{-p}D)u$$

so that

$$u = S_0 u + \sum_{p \geq 0} u_p.$$

This is the Littlewood–Paley decomposition of u. The partial sums will be denoted by $S_p u = \sum_{q=-1}^{p-1} u_q$. Each term is holomorphic in $\mathbb{C}^n$ (and, in fact, in $H^{+\infty}$ whenever u is in H^s) and the 'quality' of u is reflected in the rate of convergence of $\sum u_p$. Two lemmas will be constantly used. Let us recall the notation: $|u|_s$ denotes the norm of u in H^s $(s \in \mathbb{R})$, $\|u\|_0$ the L^∞ norm.

Lemma 1.1.1 (Almost-orthogonality of the terms). *We have*

$$1/2 \leq \psi^2(\xi) + \sum_{p \geq 0} \varphi^2(2^{-p}\xi) \leq 1, \tag{1.1.1}$$

and for all $u \in L^2$

$$\sum_{p \geq -1} |u_p|_0^2 \leq |u|_0^2 \leq 2 \sum_{p \geq -1}' |u_p|_0^2. \tag{1.1.2}$$

Proof.

$$\psi^2(\xi) + \sum \varphi^2(2^{-p}\xi) \leq \Big[\psi(\xi) + \sum \varphi(2^{-p}\xi)\Big]^2 = 1,$$

and

$$1 = \Big[\psi(\xi) + \sum \varphi(2^{-p}\xi)\Big]^2 \leq \Big(\psi^2(\xi) + \sum \varphi^2(2^{-p}\xi)\Big)$$

by virtue of the inequality $(a+b)^2 \leq 2(a^2+b^2)$. Since

$$|u|_0^2 = \text{Const.} \int |\hat{u}(\xi)|^2 \mathrm{d}\xi,$$

in order to obtain (1.1.2), it suffices to multiply the two members of (1.1.1) by $|\hat{u}(\xi)|^2$ and to integrate, remembering that $\hat{u}_p(\xi) = \varphi(2^{-p}\xi)\hat{u}(\xi)$ and $\hat{u}_{-1}(\xi) = \psi(\xi)\hat{u}(\xi)$, by definition. □

This lemma results from the fact that $(u_p, u_q)_{L^2} = 0$ whenever $|p-q| \geq 2$: the terms u_p are not orthogonal, but almost, and Lemma 1.1.1 is just Pythagoras's theorem corresponding to this almost orthogonality.

More generally, if $\{u_p\}$ is a sequence of L^2 functions with $\operatorname{supp}\hat{u}_p \subset \{\xi, \frac{1}{C}2^{p-1} \leq |\xi| \leq C2^{p+1}\}$, we always have the inequality

$$\left|\sum u_p\right|_0^2 \leq \text{Const.} \sum |u_p|_0^2. \tag{1.1.2'}$$

In what follows, given a function u on $\mathbb{R}^n$, we shall denote its L^∞ norm by $\|u\|_0$.

Lemma 1.1.2 (Sensitivity of terms to derivations). *There exist constants C, independent of p and of u, for which:*

i) *For all $\alpha \in \mathbb{N}^n$, $p \geq -1$*

$$|\partial^\alpha u_p|_0 \leq C2^{p|\alpha|}|u|_0, \quad |\partial^\alpha S_p u|_0 \leq C2^{p|\alpha|}|u|_0, \tag{1.1.3}$$

$$\|\partial^\alpha u_p\|_0 \leq C2^{p|\alpha|}\|u\|_0, \quad \|\partial^\alpha S_p u\|_0 \leq C2^{p|\alpha|}\|u\|_0. \tag{1.1.4}$$

ii) *For all $s \in \mathbb{R}$ and $p \geq 0$,*

$$\frac{1}{C}2^{ps}|u_p|_0 \leq |u_p|_s \leq C2^{ps}|u_p|_0. \tag{1.1.5}$$

iii) *For all $k \in \mathbb{N}$ and $p > 0$,*

$$\frac{1}{C}2^{pk}\|u_p\|_0 \leq \sum_{|\alpha|=k} \|\partial^\alpha u_p\|_0 \leq C2^{pk}\|u_p\|_0. \tag{1.1.6}$$

Proof. a) By definition

$$|u_p|_s^2 = (2\pi)^{-n} \int (1+|\xi|^2)^s \varphi^2(2^{-p}\xi)|\hat{u}(\xi)|^2 \mathrm{d}\xi;$$

since $(1+|\xi|^2)^s$ is bounded above and below on the support of $\varphi(2^{-p}\xi)$ by Const. $\times\, 2^{2ps}$, we immediately obtain (1.1.3) and (1.1.5).

b) Denoting the inverse Fourier transform of Φ by $\check{\Phi}$ (that is, $\hat{\check{\Phi}} = \Phi$), we have $\Phi(D)u = \check{\Phi} * u$, and if $\Phi(\xi) = \Phi_1(\mu\xi)$ ($\mu \in \mathbb{R}$; Φ_1 fixed), $\check{\Phi}(x) = \mu^{-n}\check{\Phi}_1(x/\mu)$, so that

$$\int_{\mathbb{R}^n} |\check{\Phi}(x)|\mathrm{d}x = \int_{\mathbb{R}^n} |\check{\Phi}_1(y)|\mathrm{d}y$$

is independent of μ. We deduce that $\|\check{\Phi} * u\|_0 \leq C\|u\|_0$, where C does not depend on μ. For all α, we have

$$\partial^\alpha(\Phi(D)u) = (\partial^\alpha)\check{\Phi} * u = \mu^{-|\alpha|}(\mu^{-n}(\partial^\alpha\check{\Phi}_1)(x/\mu)) * u,$$

whence

$$\|\partial^\alpha(\Phi(D)u)\|_0 \leq C\mu^{-|\alpha|}\|u\|_0;$$

applying this to $\Phi(\xi) = \varphi(2^{-p}\xi)$ (which defines u_p) or to $\Phi(\xi) = \psi(2^{-p}\xi)$ (which defines $S_p u$), we obtain (1.1.4). Similarly,

$$\begin{aligned}\partial^\alpha \hat{u}_p &= \text{Const.}\, \xi^\alpha \varphi(2^{-p}\xi)\hat{u}(\xi)\\ &= \text{Const.}\, 2^{p|\alpha|}(2^{-p}\xi)^\alpha \varphi(2^{-p}\xi)\hat{u}(\xi)\\ &= \text{Const.}\, 2^{p|\alpha|}\Phi_1(2^{-p}\xi)\hat{u}_p(\xi),\end{aligned}$$

where $\Phi_1(\xi) = \xi^\alpha \chi(\xi)$, $\chi \in C_0^\infty(\mathbb{R}^n)$, $\chi = 1$ near the support of φ. We thus obtain the right-hand inequality of (1.1.6).

c) To obtain (1.1.6), we write, for $\chi \in C_0^\infty$ equal to 1 near $\operatorname{supp}\varphi$,

$$\varphi(\xi) = \left(\sum_{|\alpha|=k} \xi^\alpha \chi_\alpha(\xi)\right)\varphi(\xi),$$

where

$$\chi_\alpha(\xi) = \frac{\xi^\alpha \chi(\xi)}{\sum_{|\alpha|=k}(\xi^\alpha)^2} \in C_0^\infty,$$

and we obtain

$$\begin{aligned}\hat{u}_p(\xi) &= \varphi(2^{-p}\xi)\hat{u}(\xi) = \sum (2^{-p}\xi)^\alpha \chi_\alpha(2^{-p}\xi)\hat{u}_p(\xi)\\ &= 2^{-pk}\sum \chi_\alpha(2^{-p}\xi)\widehat{D^\alpha u_p}(\xi),\end{aligned}$$

whence $2^{pk}u_p = \sum(2^{pn}\check{\chi}_\alpha(2^p\cdot)) * D^\alpha u_p$ and the conclusion follows. □

1.2. *Characterization of Sobolev spaces.*

Proposition 1.2. i) *If $u \in H^s(\mathbb{R}^n)$, then for all $p \geq -1$,*

$$|u_p|_0 \leq \text{Const.}\, |u|_s c_p 2^{-ps},$$

where $c_p = c_p(u)$ satisfies $\sum c_p^2 \leq 1$.

ii) *Conversely, if, for $p \geq -1$, $|u_p|_0 \leq C c_p 2^{-ps}$, with $\sum c_p^2 \leq 1$, then $u \in H^s$ and $|u|_s \leq \text{Const.}\, C$.*

Proof. Since $(\langle D\rangle^s u)_p = \langle D\rangle^s u_p$, this is an immediate consequence of (1.1.5) and of the characterization of L^2 provided by Lemma 1.1.1. □

1.3. *Characterization of Hölder spaces.* For $\alpha > 0$, $\alpha \notin \mathbb{N}$, we define $C^\alpha = C^\alpha(\mathbb{R}^n)$ to be the space of $u \in C^k(\mathbb{R}^n)$ ($k = E(\alpha)$ = integer part of α), which are bounded together with their derivatives up to order k, such that

$$\exists C, \forall x, \forall y, \forall \beta, |\beta| = k,$$

$$|\partial^\beta u(x) - \partial^\beta u(y)| \leq \mathrm{C}|x-y|^{\alpha-k}. \tag{1.3.1}$$

The norm in C^α will be denoted by $\|u\|_\alpha$ and defined by $\|u\|_\alpha = \|u\|_0 + \|u\|'_\alpha$, where $\|u\|'_\alpha$ denotes the best constant in (1.3.1).

Proposition 1.3. i) *If* $u \in C^\alpha(\mathbb{R}^n)$ $(\alpha \notin \mathbb{N})$, *then for all* $p \geq -1$,

$$\|u_p\|_0 \leq \text{Const.}\, \|u\|_\alpha 2^{-p\alpha}.$$

ii) *Conversely, if, for* $p \geq -1$, $\|u_p\|_0 \leq C2^{-p\alpha}$ $(\alpha \notin \mathbb{N})$, *then* $u \in C^\alpha$ *and* $\|u\|_\alpha \leq \text{Const.}\, C$.

Proof. We remark that the inequalities (1.1.6) and the fact that $(\partial^\alpha u)_p = \partial^\alpha u_p$ permit an immediate reduction to the case $0 < \alpha < 1$.

a) Since $\|u_{-1}\|_0 \leq \text{Const.}\, \|u\|_0$, it suffices to consider

$$u_p(x) = \int 2^{pn} \check{\varphi}(2^p(x-y))u(y)\mathrm{d}y \text{ for } p \geq 0.$$

From the fact that $\int \check{\varphi}(z)\mathrm{d}z = \text{Const.}\, \varphi(0) = 0$, we also have

$$u_p(x) = \int 2^{pn} \check{\varphi}(2^p(x-y))(u(y) - u(x))\mathrm{d}y,$$

whence

$$\begin{aligned} |u_p(x)| &\leq \|u\|_\alpha \int 2^{pn} |\check{\varphi}|(2^p(x-y))|x-y|^\alpha \mathrm{d}y \\ &\leq \text{Const.}\, \|u\|y_\alpha 2^{-p\alpha}, \end{aligned}$$

which proves i).

b) Conversely, for some p to be determined, set

$$u = S_pu + R_pu, \quad R_pu = \sum_{q \geq p} u_q;$$

then we have

$$\|R_pu\|_0 \leq \sum_{q \geq p} \|u_p\|_0 \leq C2^{-p\alpha}.$$

Moreover,

$$|S_pu(x) - S_pu(y)| \leq |x-y| \sum_{q=-1}^{p-1} \|\nabla u_q\|_0;$$

following (1.1.6),

$$\|\nabla u_q\|_0 \leq \text{Const.}\, C2^{q(1-\alpha)},$$

and of course $\|\nabla u_{-1}\|_0 \leq \text{Const.}\, C$; hence, if $0 < \alpha < 1$,

$$|S_pu(x) - S_pu(y)| \leq \text{Const.}\, C|x-y|2^{p(1-\alpha)},$$

because the series $\sum \|\nabla u_q\|_0$ is then geometrically divergent.

Regrouping the estimates for R_pu and S_pu, we find

$$|u(x) - u(y)| \leq \text{Const.}\, C|x-y|2^{p(1-\alpha)} + 2C2^{-p\alpha}.$$

If we take p to be the largest integer such that $2^p \leq \frac{1}{|x-y|}$, we obtain

$$|u(x) - u(y)| \leq \text{Const.}\, C|x - y|^\alpha,$$

which proves ii). □

Of course, one should be aware that $\|u_p\|_0 \leq C$ does not characterize L^∞ and that $\|u_p\|_0 \leq C2^{-p}$ does not characterize C^1 (in the classical sense), but a larger space (see Exercise A.3).

1.4. *Sobolev injections.*

Proposition 1.4. *For $s > n/2$, $s - n/2 \notin \mathbb{N}$, $H^s \subset C^{s-n/2}$ (continuous injection).*

Proof. We write $u_p(x) = (2\pi)^{-n} \int e^{ix\xi}\hat{u}_p(\xi)d\xi$, whence

$$\begin{aligned}\|u_p\|_0 &\leq \text{Const.} \int_{|\xi|\leq C2^p} |\hat{u}_p(\xi)|d\xi \\ &\leq \text{Const.}\, |\hat{u}_p|_0[\text{Volume}B(0, C2^p)]^{1/2} \\ &\leq \text{Const.}\, 2^{pn/2}|u|_s c_p 2^{-ps},\end{aligned}$$

following Proposition 1.2.

'Forgetting' $c_p \leq 1$, the result follows using Proposition 1.3. □

1.5. *Convexity inequalities.*

Proposition 1.5. i) *If $s = \lambda s_0 + (1-\lambda)s_1$ ($0 \leq \lambda \leq 1, s_0 < s_1, s_0, s_1 \in \mathbb{R}$), then for all $u \in C_0^\infty(\mathbb{R}^n)$, we have the inequality*

(1.5.1) $$|u|_s \leq \text{Const.}\, |u|_{s_0}^\lambda |u|_{s_1}^{1-\lambda}.$$

ii) *If $\alpha = \lambda\alpha_0 + (1-\lambda)\alpha_1$ ($0 \leq \lambda \leq 1$, $\alpha_0 < \alpha_1$, α_0, α_1 positive and not in $\mathbb{N}$), then for all $u \in C^{\alpha_1}$ we have*

(1.5.2) $$\|u\|_\alpha \leq \text{Const.}\, \|u\|_{\alpha_0}^\lambda \|u\|_{\alpha_1}^{1-\lambda}.$$

Proof. a)

$$\begin{aligned}|u|_s^2 &= \int (1 + |\xi|^2)^s |\hat{u}(\xi)|^2 d\xi \\ &= \int (1 + |\xi|^2)^{\lambda s_0} |\hat{u}(\xi)|^{2\lambda} (1 + |\xi|^2)^{(1-\lambda)s_1} |\hat{u}(\xi)|^{2(1-\lambda)} d\xi,\end{aligned}$$

and the result follows using Hölder's inequality with '$p = 1/\lambda$' and '$q = 1/1-\lambda$'.

b) We write

$$\begin{aligned}\|u_p\|_0 \leq \|u_p\|_0^\lambda \|u_p\|_0^{1-\lambda} &\leq \text{Const.}\, \|u\|_{\alpha_0}^\lambda (2^{-p\alpha_0})^\lambda \|u\|_{\alpha_1}^{1-\lambda} (2^{-p\alpha_1})^{1-\lambda} \\ &\leq \text{Const.} \|u\|_{\alpha_0}^\lambda \|u\|_{\alpha_1}^{1-\lambda} 2^{-p\alpha},\end{aligned}$$

and the result now follows by Proposition 1.3. □

Remark It is useful to know that (1.5.2) is still true for $\alpha_0, \alpha_1 \geq 0$, by defining C^α to be L^∞ for $\alpha = 0$, and the space of Lipschitz functions of order $\leq \alpha - 1$ for $\alpha \in \mathbb{N}$, $\alpha \geq 1$ (see Exercise I.2.4 a)).

Here we see very well how the characterization given in Proposition 1.3 works: in its very terms it contains a number of useful properties of Hölder spaces (for example (1.5.2)) which one does not need to know explicitly.

1.6. *Regularization operators.*

Proposition 1.6. *There exists a family S_θ ($\theta \geq 1$) of operators $S_\theta : \bigcup_{\alpha\geq 0} C^\alpha \to \bigcap_{\beta \geq 0} C^\beta$ with the following properties:*

i) $\|S_\theta u\|_\alpha \leq \text{Const.}\|u\|_\beta, \alpha \leq \beta,$

ii) $\|S_\theta u\|_\alpha \leq \text{Const.}\theta^{\alpha-\beta}\|u\|_\beta, \alpha \geq \beta,$

iii) $\|S_\theta u - u\|_\alpha \leq \text{Const.}\theta^{\alpha-\beta}\|u\|_\beta, \alpha \leq \beta,$

iv) $\|\frac{\mathrm{d}}{\mathrm{d}\theta} S_\theta u\|_\alpha \leq \text{Const.}\theta^{\alpha-\beta-1}\|u\|_\beta, \forall\alpha, \beta.$

When α or β is an integer, here we understand C^α to be the space defined by the characterization of Proposition 1.3 with the corresponding norm.

Proof. For $\chi \in C_0^\infty(\mathbb{R})$, $\chi = 1$ in the neighbourhood of the origin, we set

$$S_\theta u = \sum_p \chi(2^p/\theta) u_p,$$

implying that $(S_\theta u_p) = 0$ for $2^p \geq \text{Const.}\theta$ and $\|(S_\theta u)_p\|_0 \leq \text{Const.}\|u\|_\beta 2^{-p\beta}$ otherwise.

In particular, if $\alpha \geq \beta$,

$$\|(S_\theta u)_p\|_0 \leq \text{Const.}2^{-p\alpha}2^{p(\alpha-\beta)}\|u\|_\beta$$

implies ii).

Since $S_\theta u - u = \sum(\chi(2^p/\theta) - 1)u_p$, we have $(S_\theta u - u)_p = 0$ for $2^p \leq \text{Const.}\theta$ and $\|(S_\theta u - u)_p\|_0 \leq \text{Const.}\ \|u\|_\beta 2^{-p\beta}$ otherwise; writing

$$\|(S_\theta u - u)_p\|_0 \leq \text{Const.}\|u\|_\beta 2^{-p\alpha}2^{p(\alpha-\beta)},$$

we obtain iii).

Finally, $\frac{\mathrm{d}}{\mathrm{d}\theta} S_\theta u = \frac{1}{\theta}\sum(\chi_1)(2^p/\theta)u_p$ with $\chi_1(z) = -z\chi'(z)$, and the same arguments as before give iv), because $\text{Const.}\theta \leq 2^p \leq \text{Const.}\theta$ on the support of χ_1. □

2. Application to the study of products and composition.

2.1. Estimates of a product of two functions.

Proposition 2.1.1. *i) If $u, v \in C^\alpha$ $(\alpha \notin \mathbb{N})$,*

$$\|uv\|_\alpha \leq \text{Const.}(\|u\|_0\|v\|_\alpha + \|u\|_\alpha\|v\|_0). \tag{2.1.1}$$

ii) If $u, v \in L^\infty \cap H^2$ $(s > 0)$, then so also is uv, and

$$|uv|_s \leq \text{Const.}(\|u\|_0|v|_s + |u|_s\|v\|_0). \tag{2.1.2}$$

Proof. Let us write

$$u = \sum_p u_p, \quad v = \sum_q v_q,$$

and

$$uv = \sum_{p,q} u_p v_q = \sum_q (S_q u)v_q + \sum_p u_p S_{p+1}v = \Sigma_1 + \Sigma_2.$$

The terms Σ_1 and Σ_2 have their spectra (that is to say the supports of their Fourier transforms) in the ball $\{|\xi| \leq \text{Const}.2^p\}$. We shall use the following lemma.

Lemma 2.1. *Let $(a_q)_{q\geq -1}$ be a sequence of functions such that*

$$\operatorname{supp} \hat{a}_q \subset \{\xi, |\xi| \leq \text{Const}.2^q\}.$$

Suppose that $\|a_q\|_0 \leq C2^{-q\alpha}$ for some $\alpha > 0$ (resp. $|a_q|_0 \leq Cc_q 2^{-qs}$ for some $s > 0$, with $\sum c_q^2 \leq 1$).

Then $u = \sum_{q\geq -1} a_q$ belongs to C^α (resp. $u \in H^s$), with $\|u\|_\alpha \leq \text{Const}.C$ (resp. $|u|_s \leq \text{Const}.C$).

Proof. It suffices to observe that for some N, the dyadic blocks u_p of u satisfy $u_p = \sum_{q\geq p-N}(a_q)_p$, where, denoting the L^2 or L^∞ norm by $|\ |$,

$$|u_p| \leq \sum_{q\geq p-N} |(a_q)_p| \leq \text{Const.} \sum_{q\geq p-N} |a_q|,$$

following Lemma 1.1.2 i).

In the case of L^∞, we deduce

$$\|u_p\|_0 \leq \text{Const}.C \sum_{q\geq p-N} 2^{-q\alpha} \leq \text{Const}.C2^{-p\alpha}.$$

In the case of L^2,

$$\begin{aligned} |u_p|_0 &\leq \text{Const.}C \sum_{1\geq p-N} c_q 2^{-qs} \\ &\leq \text{Const.}C 2^{-ps} \left(\sum_{q\geq p-N} c_2^2 2^{-(q-p)s} \right)^{1/2}, \end{aligned}$$

following Schwarz's inequality applied to $c_q 2^{-(q-p)s/2} \times 2^{-(q+p)s/2}$, and it remains to remark that

$$\sum_p \sum_{q\geq p-N} c_q^2 2^{-(q-p)s} \leq \text{Const.}$$

□

We note the crucial role played in these estimates by the geometric nature of the partitioning used.

Taking into account the lemma, it suffices to evaluate $\|(S_q u)v_q\|_0$. Now

$$\|(S_q u)v_q\|_0 \leq \|S_q u\|_0 \|v_q\|_0 \leq \text{Const.}\|u\|_0 \|v\|_\alpha 2^{-q\alpha},$$

whence $\|\Sigma_1\|_\alpha \leq \text{Const.}\|u\|_0\|v\|_\alpha$, and similarly for Σ_2 with u and v interchanged.

To prove ii), we write

$$|(S_q u)v_q|_0 \leq \|S_q u\|_0 |v_q|_0 \leq \text{Const.}\|u\|_0 |v|_s c_q 2^{-qs},$$

whence $|\Sigma_1|_s \leq \text{Const.}\|u\|_0|v|_s$ and similarly for Σ_2.

If, in the above proof, we replace $uv = \sum_{p,q} u_p v_q = \Sigma_1 + \Sigma_2$ by the finer

$$\begin{aligned} uv &= \sum_{p\leq q-3} u_p v_q + \sum_{q\leq p-3} u_p v_q + \sum_{|p-q|\leq 2} u_p v_q \\ &= \sum (S_{q-2}u)v_q + \sum (S_{p-2}v)u_p + \sum_{|p-q|\leq 2} u_p v_q = \Sigma_1' + \Sigma_2' + \Sigma_3', \end{aligned}$$

we remark that the terms of the sums Σ_1' and Σ_2' have their spectra in *shells* (and not only in balls); for example, for Σ_1',

$$\begin{aligned} \text{Spectrum}\, S_{q-2}u &\subset \{\xi, |\xi| \leq 2^{q-2}\}, \\ \text{Spectrum}\, v_q &\subset \left\{\xi, \frac{1}{2}2^q \leq |\xi| \leq 2\cdot 2^q\right\}, \end{aligned}$$

whence

$$\text{Spectrum}\,(S_{q-2}u)v_q \subset \text{Sum of spectra} \subset \{\xi, 2^{q-2} \leq |\xi| \leq 9\cdot 2^{q-2}\}.$$

Moreover, the terms u_p and v_q are both small if u and v are sufficiently regular, so that the last sum, whose terms are small like $u_p v_q$, represents an 'error term' which is more regular than u or v. □

This approach is the starting point for J.-M. Bony's introduction of the concept of the 'paraproduct', where the paraproduct (nonsymmetric) of u and v is $T_u v = \sum (S_{q-2}u)v_q$ (see Exercise A.5); we then have $uv = T_u v + T_v u +$ better remainder. This concept and the associated notion of 'paradifferential operators' have turned out to be very fruitful in the study of singularities of solutions of nonlinear equations (cf. Bony [B1]).

We remark that if $\alpha \in \mathbb{N}$, part i) of the proposition is elementary, taking into account the remark of Section 1.5. A convenient version in applications is given by the following proposition.

Proposition 2.1.2. *If $u, v \in L^\infty \cap H^s$ ($s > 0$ integer), then for all α, β, with $|\alpha| + |\beta| = s$, we have*

$$|(\partial^\alpha u)(\partial^\beta v)|_0 \leq \text{Const.}(\|u\|_0 |v|_s + |u|_s \|v\|_0). \tag{2.1.3}$$

Proof. The proposition is self-evident if $|\alpha| = 0$ or $|\beta| = 0$. Otherwise (for example, $|\alpha| \geq 1$), we write

$$\partial^\alpha u \partial^\beta v = \sum_j * \partial_{i_j}(\partial^{\alpha_j} u \partial^{\beta_j} v) + * u \partial^{\alpha+\beta} v,$$

where $|\alpha_j| + |\beta_j| = s - 1$ and $*$ denotes coefficients of no importance.

To prove (2.1.3), it suffices to show that

$$|\partial^{\alpha_j} u \partial^{\beta_j} v|_1 \leq \text{Const.}(\|u\|_0 |v|_s + |u|_s \|v\|_0).$$

We then proceed as in the proof of Proposition 2.1.1:

$$\partial^{\alpha_j} u \partial^{\beta_j} v = \sum (S_q \partial^{\alpha_j} u)(\partial^{\beta_j} v)_q + \sum (\partial^{\alpha_j} u)_p (S_{p+1} \partial^{\beta_j} v) = \Sigma_1 + \Sigma_2,$$

and this time we have

$$\begin{aligned} |S_q(\partial^{\alpha_j} u)(\partial^{\beta_j} v)_q|_0 &\leq \|S_q \partial^{\alpha_j} u\|_0 |(\partial^{\beta_j} v)_q|_0 \\ &\leq \text{Const.}\|u\|_0 2^{q|\alpha_j|} |v|_s c_q 2^{-q(s-|\beta_j|)} \\ &\leq \text{Const.}\|u\|_0 |v|_s 2^{-q} c_q, \end{aligned}$$

because of Lemma 1.1.2 iii), which proves the result. □

We remark that the analogue of (2.1.3) in Hölder spaces is a simple result of the convexity inequalities of Section 1.5.

In fact, it is traditional to prove (2.1.3) using the following (so-called Gagliardo–Nirenberg) inequality (see [Au]): if $u \in L^\infty \cap H^s$ (integer $s > 0$), then for all α, $0 \leq |\alpha| \leq s$,

$$|\partial^\alpha u|_{L^p} \leq \text{Const.}\|u\|_0^{1-|\alpha|/s}|u|_s^{|\alpha|/s},$$

where $p = 2s/|\alpha|$.

Assuming this inequality, we obtain, via the Hölder inequality ($p = 2s/|\alpha|$, $q = 2s/|\beta|$)

$$\begin{aligned} |(\partial^\alpha u)(\partial^\beta v)|_0 &\leq |\partial^\alpha u|_{L^p}|\partial^\beta v|_{L^q} \\ &\leq \text{Const.}(\|u\|_0|v|_s)^{-1-|\alpha|/s}(|u|_s\|v\|_0)^{|\alpha|/s} \\ &\leq \text{Const.}(\|u\|_0|v|_s + |u|_s\|v\|_0), \end{aligned}$$

by convexity of the exponential $a^\mu b^{1-\mu} \leq \mu a + (1-\mu)b$, where the argument is parallel to that for the spaces C^α.

2.2. *Estimation of a composite function.*

Proposition 2.2. *Let $F : \mathbb{R} \to \mathbb{C}$ be a C^∞ function with $F(0) = 0$. If $u \in L^\infty \cap H^s$ ($s > 0$), then $F(u) \in L^\infty \cap H^s$ and $|F(u)|_s \leq C|u|_s$, where C depends only on F and on $\|u\|_0$.*

The proof uses the following lemma.

Lemma 2.2 (Meyer multipliers). *Let $\delta \in \mathbb{R}$, and suppose we have a sequence $m_p \in C^\infty$ with, for all $k \in \mathbb{N}$,*

$$\sum_{|\alpha|=k} \|\partial^\alpha m_p\|_0 \leq C_k 2^{p(k+\delta)}.$$

The mapping $M : u \mapsto \sum m_p u_p = Mu$ maps H^s to $H^{s-\delta}$ for all $s > \delta$, with operator norm depending only on the C_k for $k \leq E(s-\delta)+1$.

Proof. Since the spectrum of u_p is contained in $\{\xi, 1/2 \leq 2^{-p}|\xi| \leq 2\}$, let us choose $C > 4$ and partition m_p according to the formula

$$\hat{m}_p = \psi(2^{-p}\xi/C)\hat{m}_p + \sum_{k\geq 0}\varphi(2^{-k}2^{-p}\xi/C)\hat{m}_p = \hat{m}_{p,-1} + \sum_{k\geq 0}\hat{m}_{p,k}.$$

We then set $M_k u = \sum m_{p,k}u_p$, $k \geq -1$.

The terms of $M_{-1}u$ have their spectra in balls $\{\xi, |\xi| \leq (C+2)2^p\}$ and $|m_{p,-1}u_p| \leq \text{Const.}|m_{p,-1}|_0 c_p|u|_s 2^{-ps}$. Since, following Lemma 1.1.2 i),

$$\|m_{p,-1}\|_0 \leq \text{Const.}\|m_p\|_0 \leq \text{Const.}2^{p\delta}C_0,$$

Lemma 2.1 shows that $M_{-1}u \in H^{s-\delta}$ if $s > \delta$.

The terms $M_k u$ ($k \geq 0$) have their spectra in the shells

$$\left\{\xi, 2^{p+1}\left(\frac{C}{4}2^k - 1\right) \leq |\xi| \leq 2^{p+1}(1 + C2^k)\right\}$$

and

$$|m_{p,k}u_p|_0 \leq \text{Const.}\|m_{p,k}\|_0 c_p |u|_s 2^{-ps}.$$

Since, following Lemma 1.1.2 i) and iii),

$$\begin{aligned}\|m_{p,k}\|_0 &\leq \text{Const.} \sum_{|\alpha|=\ell} \|\partial^\alpha m_{p,k}\|_0 2^{-(p+k)\ell} \\ &\leq \text{Const.}C_\ell 2^{-k\ell} 2^{p\delta},\end{aligned}$$

we obtain, for all $\ell \in \mathbb{N}$,

$$\begin{aligned}|m_{p,k}u_p|_0 &\leq \text{Const.}C_\ell |u|_s 2^{-k\ell} c_p 2^{-p(s-\delta)} \\ &\leq \text{Const.}C_\ell |u|_s 2^{k(s-\ell-\delta)} C_p 2^{-(p+k)(s-\delta)}.\end{aligned}$$

We deduce that $M_k u \in H^{s-\delta}$ with $|M_k u|_{s-\delta} \leq \text{Const.}C_\ell 2^{k(s-\ell-\delta)}|u|_s$, following the remark after Lemma 1.1.1 and (1.1.2′). Finally, choosing $\ell > s-\delta$, $M = \sum_{k\geq -1} M_k$ converges normally in the space of continuous operators from H^s to $H^{s-\delta}$, and $\|M\| \leq \text{Const.}(C_0 + C_\ell)$. □

A typical example of such 'Meyer multipliers' (where $\delta = 0$) is given by $m_p = S_p a$, for some $a \in L^\infty$ (because of Lemma 1.1.2). Of course, in that case, m_p has its spectrum in a ball $\{\xi, |\xi| \leq C2^p\}$, which is not assumed in general; nevertheless, the proof of Lemma 2.2 consists precisely of showing that one can essentially reduce to that situation.

We remark (Exercise A.7) that M is a pseudo-differential operator of order δ, whose symbol $m(x,\xi) = \sum m_p(x)\varphi(2^{-p}\xi)$ does not however satisfy the standard estimates: its action from H^s to $H^{s-\delta}$ is thus not as evident as it seems, and it is therefore subject to the condition $s > \delta$.

Proof of Proposition 2.2. We use the so-called 'telescopic series' trick, which involves writing

$$F(u) = F(S_0u) + F(S_1u) - F(S_0u) + \cdots + F(S_{p+1}u) - F(S_pu) + \ldots,$$

then

$$F(S_{p+1}u) - F(S_pu) = m_p u_p, \text{ with } m_p = \int_0^1 F'(S_pu + tu_p)\mathrm{d}t.$$

a) If $u \in L^\infty \cap L^2$, for all α, $\partial^\alpha(F(S_0u)) \in L^\infty \cap L^2$: indeed

$$\partial^\alpha(F(S_0u)) = \sum * F^{(q)}(S_0u)(\partial^{\gamma_1}S_0u)\ldots(\partial^{\gamma_q}S_0u),$$

where the γ_j are multiple indices, $\gamma_1 + \cdots + \gamma_q = \alpha$, $1 \leq q \leq |\alpha|$, $|\gamma_j| \geq 1$. Each term $\partial^\gamma S_0 u$ is in $L^2 \cap L^\infty$, with $\|\partial^\gamma S_0u\|_0 \leq \text{Const.}\|u\|_0$, $|\partial^\gamma S_0u|_0 \leq \text{Const.}|u|_0$, whence the result follows if $|\alpha| \geq 1$. For $|\alpha| = 0$, $|F(S_0u)(x)| \leq C|S_0u(x)|$ (where C depends only on $\|u\|_0$), whence $|F(S_0u)|_0 \leq C\text{Const.}|u|_0$.

b) Let us verify that m_p is a 'Meyer multiplier' of order $\delta = 0$. It suffices to consider $\tilde{m}_p = G(S_p u)$. We then have

$$\partial^\alpha G(S_p u) = \sum *G^{(q)}(S_p u)(\partial^{\gamma_1} S_p u) \ldots (\partial^{\gamma_q} S_p u),$$

as in a), and $\|\partial^\gamma S_p u\|_0 \le \text{Const.}\|u\|_0 2^{p|\gamma|}$ following Lemma 1.1.2. Hence

$$\|\partial^\alpha G(S_p u)\|_0 \le \text{Const.}2^{p(|\gamma_1|+\ldots+|\gamma_q|)} = \text{Const.}2^{p|\alpha|},$$

where the constant depends only on G, α and $\|u\|_0$, which completes the proof. □

Here again, we refer readers to J.-M. Bony for the following so-called 'paralinearization' formula which makes Proposition 2.2 evident: if $u \in H^s$, $s > n/2$, $F(u) = T_{F'(u)}u + R(u)$, with T the paraproduct defined in Section 2.1 (cf. Exercise A.5), and $R(u) \in H^{2s-n/2}$ which is strictly contained in H^s. In other words, the estimate for $F(u)$ is 'linear in u' since in fact $F(u)$ is equal to $T_{F'(u)}$, up to a residue.

B. Microlocal analysis: wave front set and pseudo-differential operators

1. Wave front set of a distribution.

1.1. *Definition of the wave front set.* The Fourier transform $\hat{u}(\xi)$ of a function $u \in C_0^\infty(\mathbb{R}^n)$ is rapidly decreasing in ξ, that is,

$$\forall k, |\hat{u}(\xi)| \le C_k(1+|\xi|)^{-k}. \tag{1.1.1}$$

Conversely, if (1.1.1) is satisfied by the Fourier transform of a distribution $u \in \mathcal{E}'(\mathbb{R}^n)$, then $u \in C_C^\infty(\mathbb{R}^n)$, because of the Fourier inversion formula.

Definition 1.1.1. For $u \in \mathcal{E}'(\mathbb{R}^n)$, let $\Sigma(u)$ be the complement (in $\mathbb{R}^n \setminus 0$) of the set of directions $\xi \in \mathbb{R}^n \setminus 0$ in the neighbourhood (conical) of which $\hat{u}$ satisfies (1.1.1). By 'conical', we mean the following property of a set Γ: $\xi \in \Gamma$, $\lambda > 0 \Rightarrow \lambda\xi \in \Gamma$. Just as sing supp $u = \complement\{x,\ u \text{ is } C^\infty \text{ near } x\}$ is the set of 'bad points' of u, so $\Sigma(u)$ is the set of 'bad spectral directions' of u, or possibly 'bad frequencies' of u.

To combine these two information items in the concept of $WF(u)$, we use the following lemma.

Lemma 1.1.1. *If $\varphi \in C_0^\infty(\mathbb{R}^n)$ and $u \in \mathcal{E}'(\mathbb{R}^n)$, $\Sigma(\varphi u) \subset \Sigma(u)$.*

Proof. We have $\widehat{\varphi u}(\xi) = (2\pi)^{-n} \int \hat{\varphi}(\eta)\hat{u}(\xi - \eta)d\eta$. Since $u \in \mathcal{E}'(\mathbb{R}^n)$, $|\hat{u}(\mathcal{S})| \le C(1+|\mathcal{S}|)^M$ for some M, and, on the other hand $\hat{\varphi}$ satisfies (1.1.1). Suppose $\xi_0 \notin \Sigma(u)$: in a conical neighbourhood Γ of ξ_0, $\hat{u}$ satisfies (1.1.1);

let us split the integral into $\int_{|\eta|\leq c|\xi|}$ and $\int_{|\eta|\geq c|\xi|}$, so that for ξ in a neighbourhood Γ_1 of ξ_0, we have $\xi-\eta\in\Gamma$ in the first integral (this is possible for $0<c<1$, c sufficiently small). We then obtain

$$\left|\int_{|\eta|\leq c|\xi|}\right|\leq c_k(1+|\xi|)^{-k}(1-c)^{-k}|\hat{\varphi}|_{L_1},$$

because $|\eta|\leq c|\xi|$ implies $|\xi-\eta|\geq(1-c)|\xi|$. On the other hand

$$\begin{aligned}&\left|\int_{|\eta|\geq c|\xi|}\right|\\ \leq\ & C\int|\hat{\varphi}(\eta)|(1+|\eta|)^{k+M}(1+|\xi-\eta|)^{M}\frac{\mathrm{d}\eta}{(1+|\eta|)^{k+M}}\\ \leq\ & C\int|\hat{\varphi}(\eta)|(1+|\eta|)^{k+M}\frac{(1+(1+1/c)|\eta|)^{M}}{(1+|\eta|)^{M}}\mathrm{d}\eta\frac{1}{(1+c|\xi|)^{k}}\\ \leq\ & \frac{C(1+1/c)^{M}}{c^{k}}(1+|\xi|)^{-k}\int|\hat{\varphi}(\eta)|(1+|\eta|)^{k_M}\mathrm{d}\eta,\end{aligned}$$

because $|\xi-\eta|\leq(\frac{1}{c}+1)|\eta|$.

The estimate

$$|(1+|\xi|)^{k}\widehat{\varphi u}(\xi)\mathrm{Const.}_k\int|\hat{\varphi}(\eta)|(1+|\eta|)^{k+M}\mathrm{d}\eta \tag{1.1.2}$$

proves the lemma and at the same time makes precise the dependence upon φ. □

We can then define, for $u\in\mathcal{D}'(X)$ (X an open subset of $\mathbb{R}^n$), the set $\Sigma_x(u)$ of 'bad spectral directions' of u over x by $\Sigma_x(u)=\bigcap_\varphi\Sigma(\varphi u)$, where φ runs over the set of $\varphi\in C_0^\infty(X)$, $\varphi(x)\neq 0$.

Definition 1.1.2. Let X be an open subset of $\mathbb{R}^n$ and $u\in\mathcal{D}'(X)$. The conical closed subset of $X\times(\mathbb{R}^n\setminus 0)$ defined by

$$WF(u)=\{(x,\xi)\in\mathbb{R}^n\setminus 0,\xi\in\Sigma_x(u)\}$$

is the wave front set of u.

The following proposition shows that $WF(u)$ is the set of bad spectral directions of u over sing supp u: this is the synthesis announced in Section I.1.

Proposition 1.1.2. *The projection of $WF(u)$ onto X is* sing supp u.

Proof. In fact, if $x_0\notin$ sing supp u, $\Sigma_{x_0}(u)=\emptyset$ by definition. Conversely, let us suppose that for some x_0, $\Sigma_{x_0}(u)=\emptyset$: this means that for any direction $\xi\in S^{n-1}$, there exists $\varphi_\xi\in C_0^\infty(X)$, $\varphi_\xi(x_0)\neq 0$, such that $\widehat{\varphi_\xi u}$ is rapidly decreasing in a conical neighbourhood of ξ, V_ξ. By compactness of S^{n-1}, we obtain a finite number of functions $\varphi_i(x) = \varphi_{\xi_i}(x)$ such that, for all i, $V_{\xi_i} \cap \Sigma(\varphi_i u) = \emptyset$: Lemma 1.1.1 then implies that $\Sigma((\prod_i \varphi_i)u) = \emptyset$, that is $(\prod_i \varphi_i)u \in C_0^\infty$, which implies $x_0 \notin \text{sing supp}\, u$. □

1.2. *Examples. Case of Fourier distributions.*

Example 1.2.1. Consider δ, the Dirac mass at the origin. For all $\varphi \in C_0^\infty$, $\varphi(0) \neq 0$,

$$\widehat{\varphi\delta}(\xi) = \langle \delta, \varphi e^{-\mathrm{i}x\xi} \rangle = \varphi(0) \text{ and } \Sigma(\varphi\delta) = \mathbb{R}^n \setminus 0.$$

Thus

$$WF(\delta) = \{(x,\xi), x = 0, \xi \neq 0\}.$$

Example 1.2.2. Let

$$u = \begin{cases} -1 & x_1 < 0, \\ +1 & x_1 \geq 0. \end{cases}$$

We have

$$\widehat{\varphi u}(\xi) = -\int_{x_1 \leq 0} e^{-\mathrm{i}x\xi}\varphi(x)\mathrm{d}x + \int_{x_1 \geq 0} e^{-\mathrm{i}x\xi}\varphi(x)\mathrm{d}x.$$

When $|\xi| \to +\infty$ near a direction $\xi_0 = ((\xi_0)_1, \xi_0')$ for which $\xi_0' \neq 0$, each of the integrals is rapidly decreasing, because for example

$$\int_{x_1 \leq 0} = \int_{x_1 \leq 0} \mathrm{d}x_1 e^{-\mathrm{i}x_1\xi_1} \hat{\varphi}'(x_1, \xi')\mathrm{d}\xi'$$

where $\hat{\varphi}'$ is the partial Fourier transform in ξ'. Thus, $WF(u) \subset \{(x,\xi), x_1 = 0 \text{ and } \xi' = 0\}$.

Moreover, if a normal at $\{x_1 = 0\}$ were not in $WF(u)$, the others would not be either since u is real and translation invariant in x' and u would be C^∞. Thus $WF(u) = \{(0, x', \xi_1, 0), \xi_1 \neq 0\}$.

An important class of distributions is described in the following theorem.

Theorem 1.2. *Suppose $\varphi(x,\xi)$ is real, homogeneous of degree 1 in ξ, and C^∞ for $\xi \neq 0$, and suppose $a \in S^m$ is zero for $|x| \geq C$. If $\mathrm{d}\varphi \neq 0$ on $\operatorname{supp} a$, we can define $u = \int e^{\mathrm{i}\varphi(x,\xi)} a(x,\xi)\mathrm{d}\xi$ (oscillating integral) and $WF(u) \subset \{(x,\eta), \eta = \varphi_x'(x,\xi), \varphi_\xi' = 0\}$.*

The definition is an easy consequence of Theorem 1 of the Appendix to Chapter I (cf. Exercise I.4.6), while the control of the wave front set results from the definitions and a non-stationary phase theorem (Exercise B.9).

2. Linear operators and wave front set.

2.1. *A general theorem.* Here we shall assume the following theorem, the elementary proof of which uses only Definition 1.1.2 for the wave front set (see [H4]).

Theorem 2.1. *Let X, Y be open sets in $\mathbb{R}^n, \mathbb{R}^m$ and let $K \in \mathcal{D}'(X \times Y)$. Suppose that $WF(K)$ does not contain directions parallel to $\mathbb{R}^n$ or $\mathbb{R}^m$ (that is, $(x, y, \xi, \eta) \in WF(K) \Rightarrow \xi \neq 0,\ \eta \neq 0$). The operator K defined by the formula $Ku(x) = \int K(x,y)u(y)\mathrm{d}y$ can then be extended to $\mathcal{E}'(Y)$ (the integral being understood in the sense of distributions), and*

$$WF(Ku) \subset WF'(K) \circ WF(u), \tag{2.1.1}$$

where

$$WF'(K) = \{(x, y, \xi, \eta), (x, y, \xi, -\eta) \in WF(K)\}.$$

Here,

$$\begin{aligned} &WF'(K) \circ WF(u) \\ &= \{(x, \xi), \exists (y, \eta) \in WF(u), (x, y, \xi, \eta) \in WF'(K)\}. \end{aligned}$$

In other words, $WF'(K)$ describes the displacement of $WF(u)$ under the action of K, it being understood that (2.1.1) is only an inclusion. Here are some examples of applications of the theorem.

Example 2.1.1. Let $Tu(x) = u(x, 0)$ be the operator 'trace of u on $t = 0$', defined for $u \in C_0^\infty(\mathbb{R}^{n+1})$, where $(x, t) \in \mathbb{R}^n \times \mathbb{R}$. Since $u(x, 0) = \frac{1}{2\pi}\int \hat{u}^2(x, \tau)\mathrm{d}\tau$ ($\hat{}^2$ denotes the partial Fourier transform with respect to t), we also have

$$\begin{aligned} Tu(x) &= (2\pi)^{-n}\int e^{\mathrm{i}x\xi}\widehat{Tu}(\xi)\mathrm{d}\xi \\ &= (2\pi)^{-n-1}\int e^{\mathrm{i}x\xi}\hat{u}(\xi, \tau)\mathrm{d}\xi\mathrm{d}\tau \\ &= (2\pi)^{-n-1}\int e^{\mathrm{i}(x-y)\xi}e^{-\mathrm{i}t\tau}u(y, t)\mathrm{d}y\mathrm{d}t\mathrm{d}\xi\mathrm{d}\tau. \end{aligned}$$

The operator T thus has kernel $K(x, y, t)$ (here $X = \mathbb{R}^n$, $Y = \mathbb{R}^{n+1}$) equal to the Fourier distribution

$$K(x, y, t) = (2\pi)^{-n-1}\int e^{\mathrm{i}(x-y)\xi}e^{-\mathrm{i}t\tau}\mathrm{d}\xi\mathrm{d}\tau,$$

with phase

$$\Phi(x, y, t, \xi, \tau) = (x - y)\xi - t\tau.$$

We know (following Theorem 1.2) that $WF(K) \subset \{(x, x, 0, \xi, -\xi, -\tau)\}$. Thus, $WF(K)$ does not contain directions parallel to $\mathbb{R}^n_\xi$ but does contain directions $(\xi, -\xi, -\tau) = (0, 0, -\tau)$ parallel to $\mathbb{R}^n_{\xi,\tau}$; in order to be able to apply (2.1.1), we assume (Exercise B.4) that it is sufficient to suppose that the vertical direction $(0, -\tau)$ is not in the wave front set of u. We then find

$$WF(Tu) \subset \{(x, \xi), \exists \tau, (x, 0, \xi, \tau) \in WF(u)\}.$$

If we wish to understand this relation more intuitively, we may observe the following:

i) The singularities of u away from $\{t = 0\}$ clearly cannot play any role in Tu.

ii) If $\chi(D_x)$ is a 'tangential' operator, we have $\chi(D_\chi)Tu = T\chi(D_\chi)u$.

If for some (x_0, ξ_0) and all τ, $(x_0, 0, \xi_0, \tau) \notin WF(u)$, we will have $Pu \in C^\infty$ for $P = \tilde{\chi}(D_x, D_t)\chi(D_x)\varphi$ where $\varphi \in C_0^\infty$ is a cut-off function close to $(\chi_0, 0)$ and χ and $\tilde{\chi}$ are symbols of degree zero, with χ supported in a conical neighbourhood of x_0 and $\tilde{\chi}$ zero in a conical neighbourhood of $(0, \pm 1)$. Since the vertical is not in $WF(u)$, $\tilde{\chi}\varphi u = \varphi u + C^\infty$, whence

$$\chi(D_x)T\varphi u = \chi(D_x)\varphi(x, 0)Tu \in C^\infty,$$

which means that $(x_0, \xi_0) \notin WF(Tu)$.

Example 2.1.2. Let $X = \partial_t + \sum_{i=1}^n a_i(x, t)\partial_{x_i}$ be a C^∞ real field in $\mathbb{R}^{n+1}$. Consider the Cauchy problem

$$Xu = 0, \quad u|_{t=0} = u_0,$$

and define $Ku_0 = u(x, T)$ for $T > 0$ sufficiently small. Let $x(t, x_0)$ denote the solution of the differential equation in $\mathbb{R}^n$:

$$\begin{aligned} \frac{\mathrm{d}x_i}{\mathrm{d}t} &= a_i(x, t), \quad 1 \leq i \leq n, \\ x(0) &= x_0. \end{aligned}$$

The curve $t \mapsto (x(t, x_0), t)$ is just the integral curve of X emanating from $(x_0, 0)$; since any solution u is constant on these curves $u(x(t, x_0), t) = u_0(x_0)$, which describes the unique solution u of the Cauchy problem considered (since $(x_0, t) \mapsto (x(t, x_0), t)$ is a local diffeomorphism). Any singularity x_0 of u_0 is echoed in a singularity of u along the integral curve of X emanating from x_0, and also in a singularity of Ku_0 at $x(T, x_0)$ (because if the trace of u at $x(T, x_0)$ were C^∞, u itself would be C^∞ near $(x(T, x_0), T)$ following the preceding argument). The operator K thus displaces the singular support of u in the direction of flow of X.

Let us now examine the situation at the more detailed level of the wave front set.

We have $Ku_0(x) = u_0(\Phi^{-1}(x))$ (if $\Phi(x_0) = x(T, x_0)$), which can be written as

$$Ku_0 = (2\pi)^{-n} \int e^{\mathrm{i}(\Phi^{-1}(x)-y)\xi} u_0(y) \mathrm{d}y \mathrm{d}\xi.$$

The kernel K is thus the Fourier distribution

$$K(x,y) = (2\pi)^{-n} \int e^{\mathrm{i}(\Phi^{-1}(x)-y)\xi} \mathrm{d}\xi.$$

Following Theorem 1.2, we have (where ${}^tA$ denotes the transpose of A)

$$WF(K) \subset \{(x, \Phi^{-1}(x), {}^t(\Phi^{-1})'(x)\xi, -\xi)\},$$

and, by (2.1.1),

$$WF(Ku_0) \subset \{(\Phi(y), \xi), \xi = {}^t\Phi'^{-1}(y)\eta, (y,\eta) \in WF(u_0)\}.$$

We can 'visualize' the mapping $(y,\eta) \mapsto (x,\xi)$ induced by Φ as follows: if S is a surface passing through y with normal η at y, $\Phi(S)$ is a surface passing through x with normal ξ. At the same time, this interpretation renders intuitive the result on $WF(Ku_0)$ obtained (proceeding by analogy with Example 1.2.2). The flow of X thus induces a mapping $(y,\eta) \mapsto (x,\eta)$ which describes the displacement of the wave front set under the action of K.

Example 2.1.3. Let us consider the solution of the Cauchy problem for the wave equation

$$\text{(2.1.2)} \qquad \Box u = (\partial_t^2 - \Delta_x)u = 0, \quad u(x,0) = 0, \quad \partial_t u(x,0) = u_0$$

where $\Delta_x = \partial_1^2 + \cdots + \partial_n^2$ is the Laplacian.

Using a partial Fourier transformation in x (with abuse of notation), it is easy to calculate u; indeed

$$\partial_t^2 \hat{u} + |\xi|^2 \hat{u}(\xi,t) = 0, \quad \hat{u}(\xi,0) = 0, \quad \partial_t \hat{u}(\xi,0) = \hat{u}_0,$$

whence

$$\hat{u}(\xi,t) = A(\xi)e^{\mathrm{i}t|\xi|} + B(\xi)e^{-\mathrm{i}t|\xi|}$$

with

$$A + B = 0, \ \mathrm{i}|\xi|(A-B) = \hat{u}_0,$$

or finally,

$$u(x,t) = \frac{1}{2\mathrm{i}(2\pi)^n} \left\{ \int e^{\mathrm{i}(x-y)\xi + \mathrm{i}t|\xi|} \frac{1}{|\xi|} u_0(y) \mathrm{d}y \mathrm{d}\xi - \int e^{\mathrm{i}(x-y)\xi - \mathrm{i}t|\xi|} \frac{1}{|\xi|} u_0(y) \mathrm{d}y \mathrm{d}\xi \right\}.$$

We shall assume that u is the only solution to the problem in hand and that replacing $1/|\xi|$ by $(1-\chi(\xi))/|\xi|$ in the above integrals ($\chi \in C_0^\infty(\mathbb{R}^n)$, $\chi = 1$ near 0) defines a new function $\tilde{u}$ which differs from u only in a C^∞ function.

We remark that $\tilde{u} = (K_+ + K_-)u_0$, where the kernel K of the operator $K_\pm$ is the Fourier distribution

$$K_\pm = \frac{1}{2\mathrm{i}(2\pi)^n} \int e^{\mathrm{i}(x-y)\xi \pm \mathrm{i}t|\xi|} \frac{1-\chi(\xi)}{|\xi|} \mathrm{d}\xi.$$

Following Theorem 1.2 we have

$$WF(K_\pm) \subset \left\{ (x,t,y,\xi,\pm|\xi|,-\xi), x - y \pm t\frac{\xi}{|\xi|} = 0 \right\},$$

whence

$$WF(K_\pm u_0) \subset \left\{ (x,t,\xi,\pm|\xi|), (y,\xi) \in WF(u_0), \text{ where } y = x \pm t\frac{\xi}{|\xi|} \right\}.$$

Geometrically, let us draw Γ_y, the 'light cone' emanating from y, with equation $t^2 = |x-y|^2$: the singularity (y,ξ) of u_0 is echoed in the singularity of $K_\pm u_0$ consisting of the single direction $(\xi, \pm|\xi|)$ along the generatrix of Γ_y which is perpendicular to it.

In the particular case in which $n = 1$, $\Box = (\partial_t + \partial_x)(\partial_t - \partial_x)$, and any function $u = \Phi(x+t)$ (a solution of $(\partial_t - \partial_x)u = 0$) or $u = \Psi(x-t)$ (a solution of $(\partial_t + \partial_x)u = 0$) is a solution of $\Box u = 0$. The solution of (2.1.2) is then $u = \Phi(x-t) - \Phi(x-t)$ for $2\Phi'(x) = u_0(x)$, or $u(x,t) = 1/2 \int_{x-t}^{x+t} u_0(s)\mathrm{d}s$. As in Example 2.1.2, a singularity of u_0 will be echoed in two (since there are now two fields X) singularities of u; this analysis can be refined at the level of the wave front set.

The fundamental difference of the case $n \geq 2$ from the case $n = 1$ is that the actual position of the singular support of u depends on the wave front set of u_0 and not on its own singular support. If, for example, u_0 is C^∞ away from 0, we know that $\operatorname{sing\,supp} u \subset \Gamma_0$; but in order to know exactly which generatrices constitute $\operatorname{sing\,supp} u$, we need to know what the directions of WFu_0 are over 0. We see in this example that the introduction of the wave front set is not a simple refinement of the analysis of the singular support, it is a necessity.

2.2. *Pseudo-differentials and wave front set.* The application of the general Theorem 2.1 to pseudo-differentials is of particular interest.

Proposition 2.2.1. *Let A be a pseudo-differential operator, with kernel K. Then $WF(K) \subset \{(x,x,\xi,-\xi), \xi \neq 0\}$.*

Proof. In fact $K(x,y) = \int e^{i(x-y)\xi} a(x,\xi) d\xi$ is a Fourier distribution and the proposition follows from Theorem 1.2. □

Corollary 2.2.1. *Let A be a pseudo-differential operator. For all $u \in \mathcal{S}'(\mathbb{R}^n)$, $WF(Au) \subset WF(u)$.*

Proof. If $u \in \mathcal{E}'(\mathbb{R}^n)$, the result follows from the general Theorem 2.1 and Proposition 2.2.1, because $WF'(K) \subset \{(x, x, \xi, \xi)\}$. If $u \in \mathcal{S}'(\mathbb{R}^n)$ and $(x_0, \xi_0) \notin WF(u)$, $(x_0, \xi_0) \notin WF(\chi u)$ for $\chi \in C_0^\infty$ having value 1 near x_0. Thus $(x_0, \xi_0) \notin WF(A\chi u)$ and also $(x_0, \xi_0) \notin WF(\tilde{\chi} A \chi u)$ for $\tilde{\chi} \in C_0^\infty$ having value 1 near x_0; but $\tilde{\chi} A \chi u = \tilde{\chi} A u + C^\infty$ if $\chi = 1$ in the neighbourhood of $\operatorname{supp} \tilde{\chi}$; hence $(x_0, \xi_0) \notin WF(\tilde{\chi} A u)$ for such a choice of χ and $\tilde{\chi}$, which implies $(x_0, \xi_0) \notin WF(Au)$.

The property of Corollary 2.2.1 is called the 'pseudo-local property': it means that the action of a pseudo-differential operator does not displace the wave front set (compare with Examples 2.1.2 and 2.1.3), but, if need be, diminishes it: for example, $\partial_t \Phi(x) = 0$, while Φ can be taken to be as singular as one wishes.

We shall give more precise details of this action.

Proposition 2.2.2. *Let Γ be a conical open set in $\mathbb{R}^n \times \mathbb{R}^n$, and let $a \in S^m$, such that $a \in S^{-\infty}$ in Γ (which means that for all $(x_0, \xi_0) \in \Gamma$, there exists a conical neighbourhood of (x_0, ξ_0) on which the estimates which characterize $S^{-\infty}$ hold). Then, for $u \in \mathcal{E}'(\mathbb{R}^n)$ and $A = \operatorname{Op}(a)$, $WF(Au) \cap \Gamma = \emptyset$.*

Proof. Let $(x_0, \xi_0) \in \Gamma$ and $q(x, \xi) \in S^0$ have value 1 near (x_0, ξ_0) and be such that $aq \in S^{-\infty}$; the operator $A_1 = \operatorname{Op}(aq)$ has a C^∞ kernel, and the kernel of $A_2 = \operatorname{Op}(a(1-q))$ is C^∞ near $(x_0, x_0, \xi_0, -\xi_0)$ following Theorem 1.2. Thus $(x_0, \xi_0) \notin WF(A_2 u)$ following the general Theorem 2.1, and $Au = A_2 u + C^\infty$, which completes the proof. □

In other words, the action of an operator A destroys the wave front set where its symbol is of order $-\infty$.

The 'inverse' phenomenon is the subject of the following corollary.

Corollary 2.2.2. *If $a \in S^m$, $Au \in C^\infty$ near (x_0, ξ_0) and $|a(x, \xi)| \geq c|\xi|^m$ for $|\xi| \geq C$ in a conical neighbourhood of (x_0, ξ_0), then $(x_0, \xi_0) \notin WF(u)$.*

Proof. By definition, there exists $\tilde{b} \in S^{-m}$, $a\tilde{b} - 1 \in S^{-1}$ in a neighbourhood of (x_0, ξ_0). As in Section I.5.4, we see that there exists $b \in S^{-m}$ such that $b \# a - 1 \in S^{-\infty}$ near (x_0, ξ_0). Since $u = BAu - (BA - \mathrm{id})u$, the conclusion follows from Proposition 2.2.2 and from the pseudo-local property. □

In other words, at non-characteristic points of the symbol for A (those which, by definition, satisfy the hypothesis of Corollary 2.2.2), the wave front set is conserved by the action of A.

Corollary 2.2.2 can be used as a characterization of the wave front set: $WF(u) = \bigcap \operatorname{char}(A)$, where A describes the properly supported operators such that $Au \in C^\infty$. It is nice to have a more direct proof of Corollary 2.2.2 using only the initial definition of the wave front set. Here it is, in the case of a differential operator $P(x, D)$ in an open set X in $\mathbb{R}^n$: let $u \in \mathcal{D}'(X)$; suppose that $p_m(x_0, \xi_0) \neq 0$ and $(x_0, \xi_0) \notin WF(Pu)$.

To prove that $(x_0, \xi_0) \notin WF(u)$, we consider $\widehat{\varphi u}(\xi)$, that is to say $\langle u, \varphi e^{-ix\xi}\rangle$: the idea of the proof is to write $\varphi e^{-ix\xi}$ in the form ${}^tP(\psi e^{-ix\xi})$, such that then

$$\langle u, \varphi e^{-ix\xi}\rangle = \langle u, {}^tP(\psi e^{-ix\xi})\rangle = \langle Pu, \psi e^{-ix\xi}\rangle$$

decreases as desired according to the hypothesis on Pu. We denote $Q = {}^tP$ (with $q_m = p_m$) and observe the formula

$$e^{ix\xi} Q(\psi e^{-ix\xi}) = q_m(x, \xi)\psi + Q_{m-1}\psi + \cdots + Q_0\psi,$$

where $Q_j\psi$ is the homogeneous part in ξ of the term on the left of degree j, which is a polynomial whose coefficients are differential operators of order $m - j$ applied to ψ.

To obtain $q_m(x, \xi)\psi + \cdots + Q_0\psi = \varphi$ for (x, ξ) near (x_0, ξ_0), we simply take

$$\psi = \psi_N = \frac{1}{q_m(x, \xi)}(\varphi + a_1(x, \xi) + \cdots + a_N(x, \xi)),$$

where the $a_j(x, \xi)$ are homogeneous of degree $-j$ in ξ and are determined by the relations

$$\begin{aligned} a_1 + Q_{m-1}(\varphi/q_m) &= 0, \\ a_2 + Q_{m-1}(a_1/q_m) + Q_{m-2}(\varphi/q_m) &= 0, \text{ etc.} \end{aligned}$$

In that fashion, $e^{ix\xi}Q(\psi e^{-ix\xi}) = \varphi + r_{N+1}$, where r_{N+1} is a symbol in ξ of order $-N - 1$, $\operatorname{supp} r_{N+1} \subset \operatorname{supp} \varphi$.

What can we now say about $\langle Pu, \psi_N e^{-ix\xi}\rangle$? We know that for some $\chi \in C_0^\infty$, $\chi = 1$ near $x_0, \xi_0 \notin \Sigma(\chi Pu)$; taking $\operatorname{supp}\varphi \supset \operatorname{supp}\psi_N$ sufficiently small so that $\chi = 1$ in the neighbourhood of $\operatorname{supp}\varphi$, we obtain $\langle Pu, \psi_N e^{-ix\xi}\rangle = (\psi_N \chi Pu)(\xi)$, whence

$$\begin{aligned} (1 + |\xi|)^k |\widehat{\varphi u}(\xi)| &\leq (1 + |\xi|)^k |(\widehat{r_{N+1} u})|(\xi) \\ &\quad + C_k \int |\widehat{\psi_N}(\eta, \xi)|(1 + |\eta|)^{k+M} d\eta, \end{aligned}$$

following (1.1.2). For N sufficiently large, the second term is bounded in ξ, which completes the proof. □

Proposition 2.2.2 is also important for the following reason: if one studies an equation $Pu = f$ in the neighbourhood of a point (x_0, ξ_0), only the behaviour of the symbol p for P in the neighbourhood of (x_0, ξ_0) is involved. One can then use all the resources from the local study of functions to detail the structure of p and to deduce the properties of P.

C. Energy estimates

Here, we shall show how, in order to find and control the solutions of broad classes of operators or differential systems, it is in fact sufficient to study the case of scalar pseudo-differential operators of order 1 (cf., for example, the discussion in Example 2.1.3).

1. First-order operators. We shall consider the equation in $\mathbb{R}^{n+1}$

$$(1.1)\qquad \begin{aligned} Lu &\equiv \frac{\partial u}{\partial t} + a(t,x,D)u = f, \quad 0 < t < T,\\ u(0,x) &= \varphi, \end{aligned}$$

where

i) $a_t(x,\xi) = a(t,x,\xi)$ remains in a bounded subset of S^1 for $0 \le t \le T$,

ii) $t \mapsto a_t$ is a continuous mapping with values in C^∞,

iii) $a_1(t,x,\xi)$, the principal symbol of a, is purely imaginary.

It should be stressed that $a(t,x,D)u$ denotes the action of a as an operator, in the variables x only on $u(t,\cdot)$ (the operator is said to be 'tangential').

The prototype for the operators L is a real field $X = \partial_t + \sum a_j \partial_j$, for which $a = \mathrm{i}(\sum_j a_j \xi_j)$ satisfies i), ii) and iii). The simple fact of allowing a to be pseudo-differential, without introducing new technical difficulties, provides us with access to numerous applications.

1.1. *Energy estimate.*

Proposition 1.1. *If $s \in \mathbb{R}$ and $\lambda \in \mathbb{R}$, $\lambda \ge \lambda_s$, then for all $u \in C^1([0,T]; H^s) \cap C^0[[0,T]; H^{s+1}]$, we have the inequality*

$$(1.1.1)\qquad \sup_{t\in[0,T]} e^{-\lambda t}|u(t,\cdot)|_s \le |u(0,\cdot)|_s + 2\int_0^T e^{-\lambda t}|Lu(t,\cdot)|_s \mathrm{d}t.$$

Proof. a) Let us first suppose that $s = 0$. Then we first have

$$\begin{aligned} \frac{\mathrm{d}}{\mathrm{d}t}|\mathrm{e}^{-\lambda t}u(t,\cdot)|_0^2 &= \frac{\mathrm{d}}{\mathrm{d}t}\int \mathrm{e}^{-2\lambda t}u(t,x)\overline{u}(t,x)\mathrm{d}x\\ &= -2\lambda|\mathrm{e}^{-\lambda t}u(t,\cdot)|_0^2 + 2\mathrm{Re}\,(\mathrm{e}^{-2\lambda t}u_t, u); \end{aligned}$$

on the other hand

$$(e^{-2\lambda t}au, u) = (e^{-2\lambda t}u, a^*u) = (e^{-2\lambda t}u, (-a+b)u),$$

where $b = b(t,x,D)$, $b \in S^0$, because of iii) and Theorem I.3.2.3, whence

$$2\mathrm{Re}\,(e^{-2\lambda t}au, u) = (e^{-2\lambda t}u, bu).$$

Thus

$$2\mathrm{Re}\,(e^{-2\lambda t}Lu, u) \geq \frac{\mathrm{d}}{\mathrm{d}t}|e^{-\lambda t}u(t,\cdot)|_0^2 + (2\lambda - B)|e^{-\lambda t}u(t,\cdot)|_0^2,$$

where $B \geq \|b\|_{L^2_x \to L^2_x}$ for $0 \leq t \leq T$. For $\lambda \geq B/2$, by integrating from 0 to t, we obtain $M^2 \leq |u(0,\cdot)|_0^2 + 2MI$, with $M = \sup_{t\in[0,T]} e^{-\lambda t}|u(t,\cdot)|_0$, $I = \int_0^T e^{-\lambda t}|Lu(t,\cdot)|_0 \mathrm{d}t$, or finally,

$$M \leq |u(0,\cdot)| + 2I.$$

b) For $s \in \mathbb{R}$, we denote $E_s = (1 + |D_x|^2)^{s/2}$: the function $E_s u$ satisfies the equation $\tilde{L}E_s u = E_s Lu$, where $\tilde{L} = \partial_t + \tilde{a}$, $\tilde{a} = E_{-s}aE_s$, which has the same properties i), ii) and iii) as a. The estimate of a) then gives (1.1.1). □

The interest of this very simple proof is that it generalizes to first-order systems (of operators), that is to say, to the case in which a is an $N \times N$ matrix, when the hypothesis iii) is replaced by

iii)′ the matrix $ia_1(t,x,\xi)$ is real symmetric (or Hermitian). (See for example Exercise C.7.)

Such systems are said to be 'symmetric hyperbolic' and play a major role in the physical applications (see Section III.B.1, for example).

1.2. *Existence of solutions.* As often in the domain of partial differential equations, the existence of solutions for an operator is a consequence of inequalities first obtained for that operator.

The case considered in this section is completely typical.

Proposition 1.2. *For all $s \in \mathbb{R}$, $f \in L^1([0,T];H^s)$ and $\varphi \in H^s$, the Cauchy problem* (1.1) *has a unique solution $u \in C([0,T];H^s)$, which satisfies the energy estimate* (1.1.1).

Proof. a) Let us first verify the uniqueness. If $f = 0$, $\varphi = 0$, the equation $Lu = 0$ implies $u \in C^1([0,T];H^{s-1})$, and the inequality (1.1.1) applies (with $s-1$ instead of s), whence $u = 0$.

b) The passage from an inequality to an existence theorem uses a 'duality' method: we prove that a linear form on a certain space is continuous (this is where the inequality comes in) and we represent it using a concrete function which provides the solution.

To justify what follows, we remark that if u is a solution of (1.1), then for all $v \in C_0^\infty\{t < T\}$, we have

$$\int_0^T (u, L^*v)\mathrm{d}t = \int_0^T (Lu, v)\mathrm{d}t + (\varphi, v(0,\cdot)),$$

with $L^* = -\partial_t + a^*(t, x, D)$.

Thus we consider the antilinear form

$$\psi : L^*v \mapsto \int_0^T (f, v)\mathrm{d}t + (\varphi, v(0,\cdot)) = \psi(L^*v),$$

which is well defined on the subspace $E = L^*(C_0^\infty\{t \le T\})$ of $L^1([0,T]; H^{-s})$ because L^* satisfies the estimate (1.1.1)

$$\sup_{t\in[0,T]} |v(t,\cdot)|_{-s} \le C \int_0^T |L^*v(t,\cdot)|_{-s}\mathrm{d}t.$$

The form ψ is continuous, because

$$|\psi(L^*v)| \le \text{Const.}\left(|\varphi|_s + \int_0^T |f(t,\cdot)|_s\mathrm{d}t\right)\int_0^T |L^*v(t,\cdot)|_{-s}\mathrm{d}t,$$

following the above estimate.

The Hahn–Banach theorem enables us to extend ψ to a continuous linear form $\tilde{\psi}$. We then represent $\tilde{\psi}$ by a function $u \in L^\infty([0,T]; H^s)$ (the dual of $L^1([0,T]; H^{-s})$) according to the formula $\psi(L^*v) = \int_0^T (u, L^*v)\mathrm{d}t$, which finally gives

$$\int_0^T (u, L^*v)\mathrm{d}t = \int_0^T (f, v)\mathrm{d}t + (\varphi, v(0,\cdot)) \tag{1.2.1}$$

for all $v \in C_0^\infty(t < T)$. The remainder of the proof involves exploiting (1.2.1) ingeniously.

c) First, if we take $v \in C_0^\infty(0 < t < T)$ (that is, v also zero near $t = 0$), (1.2.1) and the definition of L^* imply $Lu = f$ in the sense of $\mathcal{D}'(0 < t < T)$. If f were continuous with values in $\mathcal{S}$, we would then obtain $\frac{\partial u}{\partial t} \in L^\infty([0,T]; H^{s-1})$, using the equation, whence $u \in C([0,T]; H^{s-1})$ and $u \in C^1([0,T]; H^{s-2})$ using the equation again.

But then, for such a u, we can integrate $\int_0^T (u, L^*v)\mathrm{d}t$ by parts and (1.2.1) implies $(u(0,\cdot), v(0,\cdot)) = (\varphi, v(0,\cdot))$ for all $v \in C_0^\infty(t < T)$ and in particular $u(0,\cdot) = \varphi$.

In this stage of the proof, we have thus obtained a solution u of (1.1) given (f, φ) in $\mathcal{S}$ and u with a regularity which enables us to apply (1.1.1) to it.

d) We then proceed by approximation, using the fact that the regularity required *a priori* for u for (1.1.1) only appears in the actual terms of this inequality; it thus disappears on passage to the limit.

For f, φ given as stated, we choose

$$f_k \in \mathcal{S}, \varphi_k \in \mathcal{S}, \text{ with } \int_0^T |(f - f_k)(t, \cdot)|_s dt \to 0, \ |\varphi - \varphi_k|_s \to 0.$$

'Thinking' that f_k, φ_k have values in H^{s+2}, point c) gives a solution $u_k \in C^1([0,T]; H^s) \cap C^0([0,T]; H^{s+1})$ of (1.1).

The energy estimate (1.1.1) applied to $u_k - u_\ell$ shows that u_k is a Cauchy sequence in $C([0,T]; H^s)$, whose limit $u \in C([0,T]; H^s)$ is the solution sought. Moreover, passing to the limit in the energy estimates for the u_k, we find the estimate for u. □

In the case of first-order symmetric systems mentioned in the preceding section, the same proof applies.

1.3. *Propagation of singularities.* By way of illustration of pseudo-differential methods, we can easily obtain the singularities of $u(t, \cdot)$ for a solution of (1.1) with $f \equiv 0$ (compare with the discussion of Example B.2.1.2 in the case of a real field).

The idea of the proof is the following: if $Q = Q(t, x, D)$ commutes with L, Qu satisfies

$$LQu = 0, \ (Qu)(0, \cdot) = Q(0, x, D)\varphi.$$

If in addition $Q(0, x, D)\varphi \in C_0^\infty$, Proposition 1.2 implies that $Q(t, x, D)u \in C^\infty$ for all t, whence

$$WFu(t, \cdot) \subset \{(x, \xi), q(t, x\xi) = 0\},$$

according to Corollary B.2.2.2.

a) Of course Q will only commute approximately with L, and we shall make this more precise.

We seek $q \sim \sum_{j \geq 0} q_j$ (q_j homogeneous of degree $-j$), and we denote $a/i \sim \sum_{j \geq 0} b_j$ (b_j homogeneous of degree $1 - j$).

The principal symbol of the commutator (tangential) $[L, Q]$ is equal to $\partial_t q_0 + \{b_0, q_0\}$, while the term of order $-j$ has symbol $\partial_t + \{b_0, q_j\} + r_j$, where r_j is a known function of $q_0, \dots, q_{j-1}$.

Thus we have to successively solve the differential equations

$$\frac{\partial q_0}{\partial t} + H_{b_0} q_0 = 0, \quad q_0(0, x, \xi) = \tilde{q}_0(x, \xi), \tag{1.3.1}$$

$$\frac{\partial q_j}{\partial t} + H_{b_0} q_j + r_j = 0, \quad q_j(0, x, \xi) = \tilde{q}(x, \xi). \tag{1.3.2}$$

If we let $(x, \xi) = \chi_t(y, \eta)$ denote the solution of the system

$$\frac{\mathrm{d}x}{\mathrm{d}t} = \frac{\partial b_0}{\partial \xi}(t, x, \xi),\ \frac{\mathrm{d}\xi}{\mathrm{d}t} = -\frac{\partial b_0}{\partial x}(t, x, \xi),\ x(0) = y,\ \xi(0) = \eta,$$

it is easy to see that χ_t exists for $0 \leq t \leq T$ and is invertible (x and ξ are homogeneous in η of degrees 0 and 1). The solution of (1.3.1) is then $q_0(t, x, \xi) = \tilde{q}_0(\chi_t^{-1}(x, \xi))$ (compare with Section B.2.1), while (1.3.2) is solved by 'varying the constant'.

If we choose $q \sim \sum q_j$, we will have $q(0, x, \xi) \sim \tilde{q}$, and $[L, Q] = R(t, x, D)$ where $r(t, x, \xi)$ is continuous in t with values in $S^{-\infty}$,

b) Let us now suppose that $(x_0, \xi_0) \notin WF(\varphi)$: then there exists $\tilde{q}(x, \xi)$, such that $\tilde{q}(x_0, \xi_0) \neq 0$ and $\tilde{q}(x, D)\varphi \in C_0^\infty$. For the solution u of (1.1), we find

$$LQu \in C([0, T]; H^s),$$
$$Qu|_{t=0} \in C_0^\infty$$

for all s (Q corresponds to $\tilde{q}$ constructed in a)).

Thus, for all t, $\chi_t(x_0, \xi_0) \notin WFu(t, \cdot)$. Since we can reverse the direction of the time, we finally find the following proposition:

Proposition 1.3. *If u is a solution of (1.1) with $f \equiv 0$ and $\varphi \in \bigcup_s H^s$, then $u \in C([0, T]); \bigcup_s H^s)$ and, for $0 \leq t \leq T$, $WFu(t, \cdot) = \chi_t(WF\varphi)$, where χ_t is defined by the system of differential equations*

$$\frac{\mathrm{d}x}{\mathrm{d}t} = \frac{1}{\mathrm{i}}\frac{\partial a_1}{\partial \xi}(t, x, \xi), \quad \frac{\mathrm{d}\xi}{\mathrm{d}t} = -\frac{1}{\mathrm{i}}\frac{\partial a_1}{\partial x}(t, x, \xi),$$
$$x(0) = y, \quad \xi(0) = \eta, \quad (x, \xi) = \chi_t(y, \eta).$$

This result can be extended in various directions:

i) For t fixed, it is easy to verify that χ is canonical (that is, it preserves the 2-form σ; cf. Section I.7.2.1).

ii) The operator $K : \varphi \to u(t_0, \cdots)$ is an 'elliptic Fourier integral operator associated with the canonical transformation χ_{t_0}'. The example developed here is the historical starting point for this theory (see [H2, H4, H5]).

On the other hand, we can deduce (Exercise C.3) from Proposition 1.3 the following theorem on the propagation of singularities due to Hörmander stated here without proof.

Theorem. *Let P be a pseudo-differential operator of order m, which is properly supported on an open set $X \subset \mathbb{R}^n$ and has real principal symbol p_m. If $u \in \mathcal{D}'(X)$ satisfies $Pu \in C^\infty(X)$, then $WF(u)$ is a union of integral curves of H_{p_m} emanating from points where $p_m = 0$.*

This theorem, which allows us to separate the rays of propagation of singularities emanating from a point $x_0 \in \operatorname{sing\,supp} u$ of the basis of X according to their origins (x_0, ξ_0) in T^*X over x_0, is one of the fundamental results of microlocal analysis (compare with the discussion of Example B.2.1.3 and Exercise C.2).

2. Operators of order m**.** When $x \in \mathbb{R}$, a differential operator of order m,

$$P(x, t, D_x, D_t) = \sum_{k+\ell \leq m} a_{k,\ell} D_x^k D_t^\ell \quad (a_{0,m} \equiv 1),$$

can be 'factorized' in the form

$$\begin{aligned} P(x, t, D_x, D_t) &= (D_t - \lambda_1(t, x) D_x) \ldots (D_t - \lambda_m(t, x) D_x) \\ &\quad + \text{order } m - 1, \end{aligned}$$

where the $\lambda_j \xi$ are the roots in τ (assumed distinct) of the polynomial $p_m = \sum_{k+\ell=m} a_{k,\ell} \xi^k \tau^\ell = 0$.

When $x \in \mathbb{R}^n$ $(n \geq 2)$, the roots in τ (assumed distinct) of the equation $p_m(x, t, \xi, \tau) = 0$ are only homogeneous of degree 1 in ξ, and the corresponding factors $D_t - \lambda_j(t, x, D_x)$ are pseudo-differential operators of the form considered in Section 1. Since the roots in τ are distinct, it is easy to show (Exercise C.4) that one can in fact factorize the complete symbol for P (and not only its principal symbol); we shall only use a more elementary fact in what follows.

Thus let us consider here an operator of the form $P = P(x, t, D_x, D_t) = \sum_{j \leq m} P_j(x, t, D_x) D_t^j$, for which $P_m \equiv 1$, where P_j is of order $m - j$ with principal symbol p_j homogeneous of degree $m - j$ in ξ.

We suppose that

(H) the roots in τ of the polynomial $p = \sum p_j(x, t, \xi) \tau^j = 0$ are real ('hyperbolicity' hypothesis for P);

(S) these roots are (uniformly) simple, that is to say $|\frac{\partial p}{\partial \tau}(x, t, \xi, \tau)| \geq C|\xi|^{m-1}$ when $p = 0$. (P is then said to be 'strictly hyperbolic'.)

For such a P, we have an energy estimate analogous to (1.1.1).

Proposition 2.1. *If $s \in \mathbb{R}$, then for all $u \in \bigcap_{j=0}^{m-1} C^j([0, T], H^{s+m-1-j})$ such that $Pu \in L^1([0, T]; H^s)$, we have the inequality*

$$\begin{aligned} &\sum_{j \leq m-1} \sup_{t \in [0,T]} |D_t^h u(t, \cdot)|_{s+m-1-j} \\ &\leq C \left\{ \sum_{j \leq m-1} |D_t^j u(0, \cdot)|_{s+m-1-j} + \int_0^T |Pu(t, \cdot)|_s \mathrm{d}t \right\}. \end{aligned} \tag{2.1}$$

Proof. a) Let $\lambda_1(x,t,\xi) < \lambda_2(x,t,\xi) < \cdots < \lambda_m(x,t,\xi)$ denote the m real zeros of p. The hypothesis (S) implies $|\partial_\xi^\beta \partial_x^\alpha \lambda_j(x,t,\xi)| \le C_{\alpha,\beta}|\xi|^{-1-|\beta|}$, so that if $\chi \in C^\infty$ takes value 1 near 0, $\Lambda_j(x,t,\xi) = (1-\chi(\xi))\lambda_j(x,t,\xi) \in S^1$. Let us show that, for all j, we can find

$$Q_j = \sum_{k=0}^{m-1} Q_{jk}(x,t,D_x)D_t^k$$

(with $Q_{jk} \in S^{m-1-k}$ and $Q_{j(m-1)} = 1$) and

$$R_j(t,x,D_x) \in S^{m-1}$$

such that

$$P = (D_t - \Lambda_j(x,t,D_x))Q_j + R_j.$$

In fact, writing

$$\begin{aligned} P_jD - t^j &= D_tP_jD_t^{j-1} - \frac{1}{\mathrm{i}}(\partial_tP_j)D_tj - 1 \\ &= (D_t - \Lambda_j)P_jD_t^{j-1} + \Lambda_jP_jD_t^{j-1} - \frac{1}{\mathrm{i}}(\partial_tP_j)D_t^{j-1} \end{aligned}$$

and starting the same trick again for the last two terms, etc., we obtain the desired expression with $R_j = R_j(t,x,D_x) \in S^m$. But, for $\tau = \lambda_j$, the principal symbol $p - (\tau - \lambda_j)q_j$ for R_j, which does not depend on τ, is zero; hence $R_j \in S^{m-1}$.

b) The energy estimate (1.1.1) for the operator $D_t - \Lambda_j$ then gives (for λ sufficiently large)

$$\begin{aligned} &\sup e^{-\lambda t}|Q_ju(t,\cdot)|_s \\ &\le |Q_ju(0,\cdot)|_s + 2\int_0^T e^{-\lambda t}\{|Pu(t,\cdot)|_s + C|u(t,\cdot)|_{s+m-1}\}\mathrm{d}t. \end{aligned}$$

We show that control of the Q_ju for all the j enables us to control the left-hand term of (2.1): the principal symbol of Q_j is $q_j = \prod_{k\ne j}(\tau - \lambda_k)$, and the q_j form a basis for the space of polynomials in τ of degree $\le m-1$; in particular, for $k \le m-1$,

$$\tau^k = \sum q_j \frac{\lambda_j^k}{q_j(x,t,\xi,\lambda_j)}.$$

Setting

$$M_{kj}(t,x,\xi) = (1-\chi(\xi))\frac{\lambda_j^k(x,t,\xi)}{q_j(x,t,\xi,\lambda_j)},$$

we define $M_{kj} \in S^{k-(m-1)}$ such that

$$D_t^k u = \sum M_{kj} Q_j u + \sum_{j=0}^{m-1} R_{kj}(t, x, D) D_t^j u,$$

with $R_{kj} \in S^{k-j-1}$. We now deduce the estimate

$$\begin{aligned} &\sum_{k \le m-1} |D_t^k u(t, \cdot)|_{s+m-1-k} \\ &\le C \left\{ \sum |Q_j u(t, \cdot)|_s + \sum_{k \le m-1} |D_t^k u(t, \cdot)|_{s+m-2-k} \right\}. \end{aligned}$$

c) The 'error term'

$$\sum_{k \le m-1} |D_t^k u(t, \cdot)|_{s+m-2-k}$$

is dealt with in a similar manner.

- For $k = m-1$, $|Q_1 u - D_t^{m-1} u|_{s-1}$ is bounded above by the sum Σ' of the terms for $k \le m-2$.
- For $k \le m-2$, we use the energy estimate (1.1.1) for the operator $L = \partial_t$, namely,

$$\begin{aligned} &\sup e^{-\lambda t} |D_t^k u(t. \cdot)|_{s+m-2-k} \\ &\le |D_t^k u(0, \cdot)|_{s+m-2-k} + 2 \int_0^T e^{-\lambda t} |D_t^{k+1} u(t, \cdot)|_{s+m-2-k} \mathrm{d}t. \end{aligned}$$

d) Finally, we obtain

$$\begin{aligned} \sup e^{-\lambda t} \sum_{k \le m-1} |D_t^k u(t, \cdot)|_{s+m-1-k} \le C \Bigg\{ \sum_{k \le m-1} |D_t^k u(0, \cdot)|_{s+m-1-k} \\ + \int_0^T e^{-\lambda t} |Pu(t, \cdot)|_s \mathrm{d}t + \int_0^T e^{-\lambda t} \sum_{k \le m-1} |D_t^k u(t, \cdot)|_{s+m-1-k} \mathrm{d}t \Bigg\}. \end{aligned}$$

The usual 'Gronwall-type' argument then allows us to conclude. □

Of course, an existence theorem for solutions of the Cauchy problem

$$Pu = f, \quad D_t^j u(0, \cdot) = u_j, \quad j \le m-1,$$

can also be deduced from (2.1) as in Section 1.2, but that is a little more delicate and we shall not continue on that path.

Commentary on Chapter II

Chapter II constitutes a bridge between the 'dry' theory of Chapter I and the applications to real nonlinear problems outlined in Chapter III.

Section A, 'Nonlinear dyadic analysis', is based on the systematic use of dyadic partitions in frequency. This approach, which has already been well developed in Coifman and Meyer [CM] has proved particularly effective in the analysis of nonlinear equations: Bony [B1] contains the fundamental concepts of 'paramultiplication', 'paradifferential operators' and the corresponding 'paralinearization' formulae, while 'paracomposition' is built up and studied in Alinhac [A]. The presentation here is inspired more by Meyer [M], which also covers the case of L^p spaces.

Beyond an introduction to the theories mentioned above, in this chapter, we wish to give the first natural examples of what we shall call 'tame applications' in Chapter III (that is to say, 'likely to be controlled by estimates which are linear with respect to the large norm'): the product and the composition.

This type of estimate is crucial for solving almost any nonlinear problem and emerges as a fundamental hypothesis of the Nash–Moser theorem.

Section B, 'Microlocal analysis', presents very classical material (wave front set, microlocal elliptic regularity), which can be found, for example, in Hörmander [H4]. Short syntheses on this subject are available: the monograph by Nirenberg [N] and the article by Hörmander [H2] in *l'Enseignement mathématique*. The approach here is via conical cut-offs in frequency, intended to separate the 'good' and 'bad' spectral directions.

Sections A and B in fact correspond to two types of subdivision of the cotangent bundle naturally associated with the classical pseudo-differential calculus (that of Chapter I): subdivision according to the size of $|\xi|$ and conical subdivision based on $\xi \in S^{n-1}$.

The development of the ideas of Section B constitutes the microlocal analysis of the 1970s: hypo-ellipticity and propagation of singularities for linear operators; it is essentially the subject of Hörmander's book [H5].

Since then, Bony's paradifferential calculus and its extensions have made it possible to extend certain fundamental results (elliptic regularity, propagation of singularities) to the case of nonlinear equations.

Finally, a need has been felt, notably in certain nonlinear problems, for an even finer localization of functions (in the cotangent bundle) in the neighbourhood of their singular points: this is 2-microlocal analysis. Knowledgeable readers may consult Lebeau [L] (in the context of analytic functions) or Bony (in the context of Sobolev spaces), for example.

Section C, 'Energy estimates', is concerned with energy inequalities which are the main tool for the study of hyperbolics and the propagation of singularities. Here again, the material is borrowed from Hörmander [H5], where one can also find extensions of this technique to the study of mixed problems.

Exercises for Chapter II

Exercise A.1. a) Let $s \in \mathbb{R}$. Show that there exists $c_s > 0$ such that if $u \in H^s$,

$$\frac{1}{c_s}|u - S_p u|_s^2 \leq \sum_{q \geq p} 4^{qs}|u_q|_0^2 \leq c_s|u - S_p u|_s^2,$$

and deduce that $\bigcap_{t\in\mathbb{R}} H^t$ is dense in H^s.

b) Let α be a positive, real, non-integer number. Set $m = E(\alpha)$ (integer part of α), $\rho = \alpha - m$ (mantissa of α). Show that there exists C_α such that, if $u \in C^\infty$,

$$\frac{1}{C_\alpha}\|u - S_p u\|_\alpha \leq \sup_{q \geq p} 2^{q\alpha}\|u_q\|_0 \leq C_\alpha\|u - S_p u\|_\alpha,$$

and deduce that $\bigcap_{\gamma\in\mathbb{R}} C^\gamma$ is not dense in C^α.

In the remainder of this exercise, we propose to characterize the closure of $\bigcap_{\gamma\in\mathbb{R}} C^\gamma$ in C^α by the condition

$$(*) \qquad \sum_{|\beta|=m} \sup_{(x,h)\in\mathbb{R}^n\times\mathbb{R}^n, |h|\leq\delta} \frac{|\partial^\beta u(x+h) - \partial^\beta u(x)|}{|h|^\rho} \to 0, \text{ when } \delta \to 0.$$

c) Verify directly that any function u of C^α in the closure of $\bigcap_{\gamma\in\mathbb{R}} C^\gamma$ satisfies the condition $(*)$.

d) Returning to the proof of Proposition 1.3 i) show that, if $u \in C^\infty$,

$$\forall q \geq 0, \qquad \|u_q\|_0 C 2^{-q\alpha} \int |\tilde{\varphi}(y)| \times \sum_{|\beta|=m} \sup_{(x,h)\in\mathbb{R}^n\times\mathbb{R}^n, |h|\leq 2^{-q}|y|} \frac{|\partial^\beta u(x+h) - \partial^\beta u(x)|}{|h|^\rho} dy.$$

Deduce that, if u satisfies $(*)$, $\|u - S_p u\|_\alpha \to 0$.

Exercise A.2 (Limiting case of Sobolev injections). a) Verify that L^1 is not contained in $H^{-n/2}$. (One might suppose the contrary and show that then the Dirac mass would belong to $H^{-n/2}$.) Deduce that $H^{n/2}$ is not contained in L^∞.

b) For $k \in \mathbb{N}$, set

$$W^{k,\infty}(\mathbb{R}^n) = \{u \in L^\infty(\mathbb{R}^n), \partial^\alpha u \in L^\infty(\mathbb{R}^n), |\alpha| \leq k\}.$$

Show that if $s = \frac{n}{2} + k$, $H^s(\mathbb{R}^n)$ is not contained in $W^{k,\infty}(\mathbb{R}^n)$. (One might remark that any function $u \in H^s$ $(s \in \mathbb{R})$ can be written as

$$u = u_0 + \sum_{j=1}^{n} \partial_j u_j, \text{ where } u_1, \ldots, u_n \in H^{s_1}, \text{ and } u_0 \in \bigcap_{t \in \mathbb{R}} H^t.)$$

Exercise A.3 (∗). (Zygmund class) Let us denote

$$C^1_*(\mathbb{R}^n) = \left\{ u \in L^\infty(\mathbb{R}^n), \quad \sup_{(x,h) \in \mathbb{R}^n \times \mathbb{R}^n} \frac{|u(x+h) + u(x-h) - 2u(x)|}{|h|} < +\infty \right\}.$$

We propose to characterize C^1_* by the dyadic decompositions.

a) Show that, if $u \in C^1_*(\mathbb{R}^n)$ and if $\varphi \in C_0^\infty(\mathbb{R}^n \setminus \{0\})$ is even, then $u_p = \varphi(2^{-p}D)u$ satisfies $\sup_{p \geq 0} \|u_p\|_0 \cdot 2^p < +\infty$.

b) We no longer suppose that φ is even in a). Show that the conclusion remains true. (One might write $u_p = \varphi(2^{-p}D)\tilde{u}_p$ and $\tilde{u}_p = \tilde{\varphi}(2^{-p}D)u$, where $\tilde{\varphi}$ is suitably chosen.)

c) Let $u \in \mathcal{S}'(\mathbb{R}^n)$, and let $u = \sum_{p \geq -1} u_p$ be a dyadic decomposition of u. Suppose that

$$\sup_{p \geq -1} \|u_p\|_0 \cdot 2^p < +\infty.$$

Show, by arguing as in the proof of Proposition 1.3 ii), that $u \in C^1_*$. (One might remark that

$$|v(x+h) + v(x-h) - 2v(x)| \leq \text{Const.}|h|^2 \sup_{|\alpha|=2} \|\partial^\alpha v\|_0.)$$

d) Exhibit the generalized Sobolev injection $H^{n/2+1}(\mathbb{R}^n) \subset C^1_*(\mathbb{T}^n)$.

e) Using Exercise A.2, show that the inclusion $W^{1,\infty} \subset C^1_*$ is strict.

Exercise A.4 (Interpolation). Suppose $s_0 < s < s_1$ are three real numbers. Let $(v_a)_{q \geq 0}$ be a sequence of H^{s_1} satisfying

$$\forall q \geq 0, \quad |v_q| \leq C c_q 2^{-q(s-s_0)}; \ |v_q|_{s_1} \leq C c_q 2^{q(s_1 - s)}$$

with

$$\sum_{q \geq 0} c_q^2 \leq 1.$$

a) Show that the series $\sum_{q \geq 0} |v_q|_{s'}$ converges for all $s' < s$. It is proposed to show that $v = \sum_{q \geq 0} v_q$ belongs to H^s, with $|v|_s \leq \text{Const.} \cdot C$.

b) Show that one can write $v_q = a_q b_q$, with

$$\operatorname{supp} \hat{a}_q \subset \{\xi, |\xi| \leq 2^{q+1}\}, \quad \operatorname{supp} \hat{b}_q \subset \{\xi, |\xi| \geq 2^q\},$$
$$|a_q|_{s_0} \leq \text{Const.} \cdot Cc_q 2^{-q(s-s_0)}, \ |b_q|_{s_1} \leq \text{Const.} \cdot Cc_q 2^{q(s_1-s)}.$$

c) By estimating the L^2 norms of the dyadic blocks of $\sum_{q\geq 0} a_q$, $\sum_{q\geq 0} b_q$, show that $\sum_{q\geq 0} a_q$, $\sum_{q\geq 0} b_q \in H^s$ and conclude.

d) What is the link between this result and that of Lemma 2.2?

e) Show that the conclusion is the same if one replaces the conditions (*) by

$$\forall q \geq 0, \quad |v_q|_{s_0} \leq Cc_q A^{-q(s-s_0)}; \ |v_q|_{s_1} Cc_q A^{q(s_1-s)},$$

with $\sum_{q\geq 0} c_q^2 \leq 1$, $A > 1$. (One might define and used the A-adic decompositions.)

f) Application. Let t_0, t_1, s_0, s_1 be four real numbers, and let $T : \mathcal{S}' \to \mathcal{S}'$ be a bounded linear operator from H^{t_i} to H^{s_i} for $i = 0, 1$. Let $\theta \in]0, 1[$; set $t = \theta t_0 + (1-\theta)t_1$; $s = \theta s_0 + (1-\theta)s_1$. Show that T is bounded from H^t to H^s. (Estimate $|Tu_p|_{s_i}$, where $u = \sum u_p$ is the dyadic decomposition of u.)
Remark. Exercise A.4 can also be dealt with in the framework of Hölder spaces.

Exercise A.5. Let $a \in L^\infty(\mathbb{R}^n)$. Define the 'paraproduct by a' by the following formula:

$$T_a u = \sum_{p\geq 2} S_{p-2} a \cdot u_p$$

(with the usual notation for the dyadic decomposition; cf. Section 1.1).

a) Show that T_a is bounded on H^s for all $s \in \mathbb{R}$, and on C^α for all non-integer $\alpha > 0$, with an operator norm bounded by Const.$\|a\|_0$.

b) Let $a \in H^s(\mathbb{R}^n)$, $s < n/2$. Show that

$$\|S_p a\|_0 \leq C 2^{p((n/2)-s)},$$

and deduce that T_a, again defined according to the above formula, is continuous from H^r to $H^{r+s-n/2}$ for all $r \in \mathbb{R}$.

c) Let s, t be two real numbers such that $s + t > n/2$. Let $a \in H^s(\mathbb{R}^n)$, $b \in H^t(\mathbb{R}^n)$. Show that $ab = T_a b + T_b a + R$ with $R \in H^{s+t-n/2}(\mathbb{R}^n)$ (one might use Lemma 2.1). Discuss the H^r regularity of the product ab according to the values of $s, t, s+t$.

Exercise A.6. Suppose $f \in C^\infty(\mathbb{R})$ is such that $f(0) = 0$ and, for all $k \geq 1$, $f^{(k)}$ is bounded on $\mathbb{R}$. It is proposed to prove that, if $u \in H^{n/2}(\mathbb{R}^n)$, then $f(u) \in H^{n/2}(\mathbb{R}^n)$.

a) If $u \in H^{n/2}(\mathbb{R}^n)$, prove that

$$\forall \alpha, \quad |\alpha| > 0, \; \|\partial^\alpha S_p u\|_0 \le C_\alpha 2^{p|\alpha|}.$$

Deduce that the $m_p = \int_0^1 f'(S_{p-1}u + tu_p)\mathrm{d}t$ are 'Meyer multipliers', and conclude by returning to the proof of Proposition 2.2.

b) Show similarly that for $s > n/2$, $|f(u)|_s \le C_s |u|_s$ where C_s depends only on $|u|_{n/2}$. (Compare with Proposition 2.2.)

Exercise A.7 (*). Let $(m_p)_{p\ge 0}$ be a sequence of C^∞ functions on $\mathbb{R}^n$ satisfying

$$\forall p, \forall \alpha, \quad \|\partial^\alpha m_p\|_0 \le C_\alpha 2^{p|\alpha|}.$$

Let $\varphi \in C^\infty(\mathbb{R}^n)$ be such that $\operatorname{supp}\varphi \subset \{\xi, 1/4 \le |\xi| \le 4\}$.

a) Show that the function defined by

$$a(x,\xi) = \sum_{p=0}^{\infty} m_p(x)\varphi(2^{-p}\xi)$$

is a symbol of the class $S^0_{1,1}$, i.e. it satisfies the estimates

$$|\partial_x^\alpha \partial_\xi^\beta a(x,\xi)| \le C_{\alpha,\beta}(1+|\xi|)^{|\alpha|-|\beta|}.$$

b) Let $b \in S^0_{1,1}(\mathbb{R}^n \times \mathbb{R}^n)$. Show that one can write

$$b(x,\xi) = b_{-1}(x,\xi) + \sum_{p=0}^{\infty} b_p(x; 2^{-p}\xi)$$

with

$$\forall p \ge -1,\; b_p \in C^\infty(\mathbb{R}^n \times \mathbb{R}^n), |\xi| \le 1 \text{ on } \operatorname{supp} b_{-1},$$

$$1/2 \le |\xi| \le 2 \text{ on } \operatorname{supp} b_p \text{ for } p \ge 0,$$

and

$$|\partial_x^\alpha \partial_\xi^\beta b_p(x,\xi)| \le C_{\alpha\beta} 2^{p|\alpha|}.$$

Deduce that

$$b(x,\xi) = b_{-1}(x,\xi) + \int_{\mathbb{R}^n} a(x,\xi,y)\mathrm{d}y$$

where $a(x,\xi,y) = \sum_{p=0}^\infty m_p(x,y)\Phi(2^{-p}\xi, y)$, for each y, is of the type con sidered in a) with, in addition

$$\forall p, \forall \alpha, \forall N, \qquad |\partial_x^\alpha m_p| \le C_{\alpha,N} 2^{p|\alpha|}(1+|y|)^{-N},$$
$$\forall \alpha, \qquad |\partial_\xi^\alpha \Phi| \le C_\alpha (1+|y|)^{|\alpha|}.$$

c) Show that, if $b \in S^0_{1,1}(\mathbb{R}^n \times \mathbb{R}^n)$, then $b(x,D)$ is bounded on H^s for all $s > 0$. (Reduce to Proposition 2.2 via b).)

d) Show similarly that $b(x, D)$ is bounded on C^α for all non-integer $\alpha > 0$.

Remark. The continuity of pseudo-differential operators of class $S^0_{1,1}$ on Sobolev spaces is a delicate problem; these operators are not in general bounded on L^2 (see [CM], pp. 39–40). We refer to [H8] for a recent study of this type of problem.

Exercise B.1. Let S be the hypersurface $\{x_1 = 0\}$, and let u be a function of the form

$$u(x) = u_+(x) \cdot 1_{\{x_1>0\}} + u_-(x) \cdot 1_{\{x_1<0\}},$$

where u_+, u_- are C^∞ functions on $\mathbb{R}^n$ satisfying $\forall x' \in \mathbb{R}^{n-1}, u_+(0, x') \neq u_-(0, x')$.

a) Show that $WFu = \{(0, x', \xi_1, 0), \xi_1 \neq 0, x' \in \mathbb{R}^{n-1}\}$. Using the invariance of the wave front set under diffeomorphisms, express this result intrinsically when S is an arbitrary hypersurface (compare with Example 2.1.2).

b) Let $P = \sum_{|\alpha|\leq m} a_\alpha(x)D^\alpha$ be a differential operator with C^∞ coefficients. Calculate $Pu(x)$ explicitly and deduce that if $Pu \in C^\infty$, then S is characteristic for P (i.e. $p_m(x, \xi) = 0$ for all $x \in S$, ξ normal to S at x, where p_m is the principal symbol for P). Could one predict this result?

Exercise B.2 (Construction of distributions having a given wave front set). Suppose $\varphi \in C_0^\infty(\mathbb{R}^n)$ is such that $\int \varphi(x)dx = 1$, and let $\alpha \in]0, \frac{1}{2}[$.

a) For $k \geq 1$, $\xi \in S^{n-1}$, $x \in \mathbb{R}^n$, set

$$\varphi_k(x, \xi) = k^{n\alpha}\varphi(k^\alpha x)e^{ikx\cdot\xi}.$$

Show that the Fourier transform of φ_k relative to x satisfies

i) $\forall N, |\hat{\varphi}_k(\eta, \xi)| \leq C_N(1 + |\eta|)^{-N}$ if η remains in a closed cone Γ disjoint from $\mathbb{R}_+^*\xi$, where C_N depends on Γ,

ii) $\hat{\varphi}_k(k\xi, \xi) = 1$.

b) Suppose in this question that $\hat{\varphi} \geq 0$, and let $\xi_0 \in S^{n-1}$. Show that the function

$$u(x) = \sum_{k=1}^{\infty} k^{-2n\alpha-2}\varphi_k(x, \xi_0)$$

is continuous with compact support on $\mathbb{R}^n$ and that

$$\Sigma(u) = \{\xi_0\}, \ WF(u) = \{(0, t\xi_0), t > 0\}.$$

c) Let F be a conical closed subset of $\mathbb{R}^n \times (\mathbb{R}^n \setminus 0)$; denote $F_1 = F \cap (\mathbb{R}^n \times S^{n-1})$ and suppose we have a sequence $(x_k, \xi_k)_{k\geq 1}$ which is dense in F_1. Show that the function

$$u(x) = \sum_{k=1}^{\infty} k^{-2n\alpha-2} \varphi_{k^2}(x - x_k, \xi_k)$$

is continuous and satisfies $WF(u) = F$. (One might remark that, if $k \neq j, \xi \in S^{n-1}, \eta \in S^{n-1}$, then

$$|\hat{\varphi}_{k^2}(j^2\xi, \eta)| \leq C_N j^{-N}$$

for all N; deduce that, if $\theta \in C_0^\infty(\mathbb{R}^n)$ satisfies $\theta(x) = 1$ near $x = x_j$, then

$$|\widehat{\theta u}(j^2 \xi_j)| \geq \frac{1}{2} j^{-2n\alpha-2},$$

for j sufficiently large.)

Exercise B.3. Show that if u is a real-valued distribution, $(x, \xi) \in WFu$ implies $(x, -\xi) \in WFu$.

Exercise B.4 (Microlocal study of the trace). Suppose $u \in \mathcal{D}'(\mathbb{R}^n_x \times \mathbb{R}_t)$ is such that $WFu \cap \{t = 0, \xi = 0\} = \emptyset$.

a) Prove that if $\psi \in C_0^\infty(\mathbb{R}_t)$ is supported sufficiently close to $t = 0$, then the integral

$$I(\varphi) = \int \hat{v}(0, \tau) \frac{\mathrm{d}\tau}{2\pi}, \text{ where } v = (\psi \otimes \varphi)u,$$

converges for all $\varphi \in C_0^\infty(\mathbb{R}^n_x)$, does not depend on the choice of ψ (if one also imposes $\psi = 1$ near $t = 0$) and defines a distribution Tu on $\mathbb{R}^n$.

b) Verify that, if in addition $u \in C(\mathbb{R}_+, \mathcal{D}'(\mathbb{R}^n_x))$, we have $Tu = u|_{t=0}$.

c) Use the definition of the wave front set to show that

$$WF(Tu) \subset \{(x, \xi), \exists \tau, (x, 0, \xi, \tau) \in WF(u)\}.$$

Exercise B.5. This exercise proposes a proof of Proposition 2.2.2 and Corollary 2.2.1 independently of the general theorems cited in Section B.

a) Let $a \in S^m(\mathbb{R}^n \times \mathbb{R}^n)$ and $v \in \mathcal{E}'(\mathbb{R}^n)$. Prove that if

$$\mathrm{d}(x, \operatorname{supp} v) \geq c > 0, \forall \alpha, \forall \beta, \ |x^\alpha \partial_x^\beta a(x, D)v| \leq C_{\alpha\beta}.$$

Deduce that if $a(x, D)v \in C^\infty$, then $a(x, D)v \in \mathcal{S}$.

b) Show that $(x_0, \xi_0) \notin WFu$ if and only if there exist $\varphi(x) \in C_0^\infty$ which is non-zero at x_0 and $\chi(\xi) \in S^0$ which is elliptic at ξ_0, such that $\chi(D)(\varphi u) \in C^\infty$.

c) If a is a symbol, we define the essential support of a (denoted by $ES(a)$) to be the set of $(x_0, \xi_0) \in T^*\mathbb{R}^n \setminus 0$ which do not have a conical neighbourhood Γ on which $a \in S^{-\infty}$. If $A = a(x, D)$, we again denote $ES(A) = ES(a)$. Show that $ES(AB) \subset ES(A) \cap ES(B)$.

d) First prove Proposition 2.2.2 by remarking that, if $(x_0, \xi_0) \notin ES(A)$, then there exist χ and φ such that $\chi(D)\varphi$ is elliptic in (x_0, ξ_0) and

$$ES(\chi(D)\varphi) \cap ES(A) = \emptyset.$$

e) Prove Corollary 2.2.1 by writing

$$A = A_\chi(D)\varphi + A(1 - \chi(D)\varphi),$$

where φ has value 1 near x_0 and χ has value 1 in a conical neighbourhood of ξ_0, and applying Proposition 2.2.2.

Exercise B.6. Let $s \in \mathbb{R}$. If $v \in \mathcal{E}'(\mathbb{R}^n)$, let $\Sigma^s(v)$ denote the set of $\xi \in \mathbb{R}^n\backslash 0$ which do not contain a conical neighbourhood Γ for which $(1+|\eta|^2)^{s/2}\hat{v}(\eta) \in L^2(\Gamma)$. If X is an open set in $\mathbb{R}^n, x \in X$ and $u \in \mathcal{D}'(X)$, we then define $\Sigma^s_x(u)$ and $WF_s(u)$ (H^s wave front set of u) according to the same scheme as in Section 1.1 (with an evident extension of Lemma 1.1.1).

a) Characterize the projection of $WF_s(u)$ on X,

b) Show that $WF(u)$ is equal to the closure of $\bigcup_{s\in\mathbb{R}} WF_s(u)$. (One might remark that, if $\bigcup_s \Sigma^s(v) \cap \overline{\Gamma} = \emptyset$, with $v \in \mathcal{E}'$ and Γ a conical open set, then $\Sigma(\varphi v) \cap \overline{\Gamma}_1 = \emptyset$ for all $\varphi \in C_0^\infty(\mathbb{R}^n)$ and any conical open set Γ_1 with $\overline{\Gamma}_1 \subset \Gamma$.)

c) Show that $(x_0, \xi_0) \notin WF_s(u)$ if and only if one can write $u = u_1 + u_2$ with $u_1 \in H^s_{\text{loc}}(X)$, $(x_0, \xi_0) \notin WF(u_2)$.

d) Show that, if P is a properly supported pseudo-differential operator of order m on X, then

$$WF_s(Pu) \subset WF_{s+m}(u) \subset WF_s(Pu) \cup \text{Char}(P).$$

e) Under the hypotheses of d), suppose $m > 0$. Show that if $P^k u \in L^2_{\text{loc}}(X)$ for all $k \in \mathbb{N}$, then $WF(u) \subset \text{Char}(P)$.

Exercise B.7 (*). (Rauch's Lemma: see [Ra]) Let $s > n/2$ and $u, v \in H^s_{\text{loc}}(X)$. It is proposed to show that for all $t \leq 2s - n/2$,

$$WF_t(uv) \subset WF_t(u) \cup WF_t(v).$$

a) Verify that it suffices to prove the following assertion: if $u, v \in H^s(\mathbb{R}^n)$ have compact supports and $t \leq 2s - n/2$, then $\Sigma^t(uv) \subset \Sigma^t(u) \cup \Sigma^t(v)$.

b) Assume the situation of a). Denote $(1 + |\xi|^2)^{1/2} = \langle\xi\rangle$. Let Γ be an open cone such that $\langle\xi\rangle^t u, \langle\xi\rangle^t v \in L^2(\Gamma)$. For $c \in]0, 1[$, to be chosen, write

$$\widehat{uv} = w_1 + w_2 + r, \quad w_1(\xi) = \int_{|\xi-\eta|<c|\eta|} \hat{u}(\xi - \eta)\hat{v}(\eta)\mathrm{d}\eta,$$

$$w_2(\xi) = \int_{|\xi-\eta|>\frac{1}{c}|\eta|} \hat{u}(\xi - \eta)\hat{v}(\eta)\mathrm{d}\eta.$$

Show that if Γ_1 is an open cone satisfying $\overline{\Gamma}1 \subset \Gamma$, for some appropriate c, then $\langle\xi\rangle^{t} w_i \in L^2(\Gamma_1)$ for $i = 1, 2$. Hint: show that

$$\langle\xi\rangle^{t}|w_i(\xi)| \leq C\left(\int |\hat{u}(\eta)|\mathrm{d}\eta\right)^{1/2}\left(\int_\Gamma |\hat{u}(\xi-\eta)||\hat{v}(\eta)|^2\mathrm{d}\eta\right)^{1/2}.$$

c) Remark that for positive numbers c_1, c_2,

$$|r(\xi)| \leq \left(\int_{|\eta_1|\geq c_1|\xi|} |\hat{u}(\eta_1)|^2\mathrm{d}\eta_1\right)^{1/2}\left(\int_{|\eta_2|\geq c_2|\xi|} |\hat{v}(\eta_2)|^2\mathrm{d}\eta_2\right)^{1/2}$$

and deduce that $\langle\xi\rangle^{2s-n/2}r \in L^2(\mathbb{R}^n)$ (compare with Exercise A.5 c)).

Exercise B.8 $(*)$. Let X be an open subset of $\mathbb{R}^n$, and $\mathcal{V}$ a family of C^∞ vector fields on X. Define

$$\begin{aligned} H^{s,\infty}(\mathcal{V}) \quad = \quad &\{u \in H^s_{\mathrm{loc}}(X), Z_1 \ldots Z_k u \in H^s_{\mathrm{loc}}(X) \\ &\qquad \forall k \in \mathbb{N}, \forall Z_1, \ldots, Z_k \in \mathcal{V}\}. \end{aligned}$$

Denote the Lie algebra generated by $\mathcal{V}$ by $\hat{\mathcal{V}}$ and

$$\mathcal{N}(\mathcal{V}) = \{(x,\xi) \in T^*X \setminus 0, \langle\xi, Z(x)\rangle = 0 \ \forall Z \in \hat{\mathcal{V}}\}.$$

a) Show that, if $u \in H^{s,\infty}(\mathcal{V})$, $WF(u) \subset \mathcal{N}(\mathcal{V})$.

b) Let S be a hypersurface of $\mathbb{R}^n$; take $\mathcal{V}$ to be the set of all vector fields tangent to S. Give an explicit expression for the result of a) in this case. Give examples of elements of $H^{s,\infty}(\mathcal{V})$. (Think of the 'piecewise C^∞' distributions of Exercise B.1.)

c) Suppose that $s > n/2$. Show that if $u \in H^{s,\infty}(\mathcal{V})$, then $f(u) \in H^{s,\infty}(\mathcal{V})$.

Exercise B.9 $(*)$. Let $I_\varphi(a) = \int e^{\mathrm{i}\varphi(x,\theta)}a(x,\theta)\mathrm{d}\theta$ be an oscillating integral distribution defined as in Exercise I.4.6 b).

a) Let $\psi = \psi(x)$ be a real-valued C^∞ function on Ω, such that $\mathrm{d}\psi \neq 0$ on Ω and $\forall x \in \Omega, \forall\theta \in \mathbb{R}^N \setminus 0$,

$$\mathrm{d}_\theta\varphi(x,\theta) = 0 \Rightarrow \mathrm{d}\psi(x) \neq \mathrm{d}_x\varphi(x,\theta).$$

Show that for any compact subset K of $\Omega \times \mathbb{R}^N \setminus 0$, there exists a constant $C > 0$ such that $\forall\lambda \geq 0, \forall R \geq 0, \forall(x,\theta) \in K$,

$$R|\mathrm{d}_\theta\varphi(x,\theta)| + |R\mathrm{d}_x\varphi(x,\theta) - \lambda\mathrm{d}\psi(x)| \geq C(R+\lambda).$$

Deduce that if $u \in C_0^\infty(\Omega)$,

$$|\langle I_\varphi(a), ue^{-\mathrm{i}\lambda\psi}\rangle| \leq C_N\lambda^{-N} \text{ for all } \lambda \geq 1.$$

(One might write $a = \sum_p a\chi(2^{-p}\theta)$ as in the proof of Theorem 1 of the Appendix of Chapter I and apply the non-stationary phase Lemma 1.)

b) Conclude that $WF(I_\varphi(a)) \subset \{(x, \mathrm{d}_x\varphi(x,\theta)), \mathrm{d}_\theta\varphi(x,\theta) = 0\}$.

Exercise C.1. This exercise proposes a slightly different proof for the result on the existence of solutions to a hyperbolic problem (Proposition 1.2). The 'duality' method is replaced by a method of approximation of the problem by a family of ordinary differential equations.

We assume the hypotheses of Proposition 1.2. For $\varepsilon \in]0,1[$, set $\rho_\varepsilon = (1+\varepsilon|D|)^{-1}$ and $a_\varepsilon(t,x,D) = \rho_\varepsilon a(t,x,D)$. We approximate φ in H^s and f in $L^1([0,T];H^s)$ by families (φ_ε) and (f_ε) of elements of $\mathcal{S}$.

a) Show that, for any $\varepsilon > 0$, the equation

$$\frac{\partial u_\varepsilon}{\partial t} + a_\varepsilon(t,x,D)u_\varepsilon = f_\varepsilon, \quad u_\varepsilon(0,x) = \varphi_\varepsilon(x),$$

has precisely one solution $u_\varepsilon \in C^1([0,T];H^r)$, for all $r \in \mathbb{R}$.

b) By applying the hyperbolic energy inequality, and using Exercise I.5.10, show that (u_ε) is bounded in $C([0,T];H^s)$ and is Cauchy in $C([0,T];H^{s-2})$ when ε tends to zero. Deduce the existence of a solution $u \in C([0,T];H^{s-2}) \cap L^2([0,T];H^s)$ to the initial problem.

Arguing as at the end of the proof of Proposition 1.2 and using the uniqueness (proved separately), we obtain $u \in C([0,T];H^s)$.

Exercise C.2 (Wave equation). Let $\varphi_0 \in H^s(\mathbb{R}^n)$, $\varphi_1 \in H^{s-1}(\mathbb{R}^n)$. It is proposed to solve the problem

$$(\partial_t^2 - \Delta)u = 0,\ u|_{t=0} = \varphi_0,\ \partial_t u|_{t=0} = \varphi_1, \tag{1}$$

with $u \in C(\mathbb{R},H^s) \cap C^1(\mathbb{R},H^{s-1})$, and to study the microlocal regularity of u. Let $D|$ denote the operator defined by

$$(|D|v)(\xi) = |\xi|\hat{v}(\xi) \text{ for } v \in \mathcal{S}'(\mathbb{R}^n).$$

This operator is not pseudo-differential in the strict sense of the definition in Section I.3: we leave the reader to verify that this detail is of no importance for the results below.

a) We suppose that u is a solution of (1) with the regularity stated and set $v_+ = (\partial_t + \mathrm{i}|D|)u$, $v_- = (\partial_t - \mathrm{i}|D|)u$. Show that $v_\pm \in C(\mathbb{R},H^{s-1})$ and are solutions of the system

$$(\partial_t \mp \mathrm{i}|D|)v_\pm = 0,\ v_\pm|_{t=0} = \varphi_1 \pm \mathrm{i}|D|\varphi_0, \tag{2}$$

while u satisfies

$$\partial_t u = \frac{v_+ + v_-}{2},\ u|_{t=0} = \varphi_0. \tag{3}$$

Deduce the uniqueness for the problem (1).

b) For the existence, show that one can solve the system (2) using the results of (1), then that (3) defines $u \in C^1(\mathbb{R},H^{s-1})$, satisfying in addition $i|D|u = (v_+ - v_-)/2$. Deduce that u is a solution of (1) with the stated regularity.

c) Show that, for all t,

$$WF(v_\pm(t)) = \left\{ \left(x \mp t\frac{\xi}{|\xi|}, \xi \right), (x,\xi) \in WF(\varphi_1 \pm \mathrm{i}|D|\varphi_0) \right\},$$

and prove that $WF(u(t)) \subset WF(v_+(t)) \cup WF(v_-(t))$ with equality if these two sets are disjoint. Compare with the Example B.2.1.3.

d) Show that v_+, v_- are solutions of the wave equation, and deduce (using Exercise B.4) that

$$WF(v_\pm(t)) = \left\{ \left(x \mp t\frac{\xi}{|\xi|}, t, \xi, \pm|\xi| \right), t \in \mathbb{R}, (x,\xi) \in WF(\varphi_1 \pm \mathrm{i}|D|\varphi_0) \right\}.$$

Finally, show that $WFu = WF(v_+) \cup WF(v_-)$.

Exercise C.3 (*). (Hörmander's theorem) a) Let $a = a(x,\xi) \in S^1(\mathbb{R}^n \times \mathbb{R}^n)$ be real-valued with principal symbol $a_1(x,\xi)$. Let $u \in \bigcup_{s\in\mathbb{R}} H^s(\mathbb{R}^n)$. Suppose that $f = a(x,D)u \in \bigcap_{s\in\mathbb{R}} H^s(\mathbb{R}^n)$. Remarking that the function $v(t,x) = u(x)$ $(t \in \mathbb{R})$ is a solution of

$$(\partial_t + \mathrm{i}a(x,D))v = \mathrm{i}f (\forall s, f \in C^\infty(\mathbb{R}_t, H^s)),$$

show that, if $(x,\xi) \in WFu$, then $\chi_t(x,\xi) \in WFu$.

For all $t \in \mathbb{R}$, we set

$$\chi_t(y,\eta) = (x(t),\xi(t)) \text{ with } \quad \dot{x}(t) = \frac{\partial a_1}{\partial \xi}(x(t),\xi(t)),$$
$$\dot{\xi}(t) = -\frac{\partial a_1}{\partial x}(x(t),\xi(t)),$$
$$x(0) = y, \quad \xi(0) = \eta.$$

(Verify that Proposition 1.3 remains true if one adds a second regular member.)

b) Let P be a pseudo-differential operator of order m which is properly supported on an open set X of $\mathbb{R}^n$ (or a differentiable manifold). Suppose that the principal symbol p_m is real. Let $u \in \mathcal{D}'(X)$ be such that $Pu \in C^\infty(X)$. Show that WFu is a union of bicharacteristics of p_m (i.e. integral curves of H_{p_m} associated with the value $p_m = 0$).

Hint: Reduce to a situation on all of $\mathbb{R}^n$ as in a) using cut-off functions, then to the case $m = 1$ by multiplying P by an elliptic operator of order $1 - m$.

c) Apply the result stated in b) to the case where P is a vector field $P = \partial_t^2 - \Delta$ (compare with the final result of Exercise C.2).

d) Consider $P = \partial_x^2 + x\partial_y^2$ on the band $B = \{-1 < x < 1, y \in \mathbb{R}\}$. Let $u \in \mathcal{D}'(B)$ be such that $Pu \in C^\infty(B)$. Suppose that $u \in C^\infty$ near to $\{x = -c\}$ for some $c \in]0,1[$. Show that $u \in C^\infty(B)$.

Exercise C.4. Let $P(x,t,D_x,D_t) = \sum_{j=0}^{\infty} P_j(t,x,D_x)D_t^j$ be a differential operator of order m. Denote its principal symbol by

$$p(x,t,\xi,\tau) = \sum_{j=0}^{m} p_j(t,x,\xi)\tau^j;$$

suppose that $p_m = 1$ and that the roots relative to τ of the equation $p(x,t,\xi,\tau) = 0$ are distinct (denoted by $\tau_1(t,x,\xi),\dots,\tau_m(t,x,\xi)$).

It is proposed to show that there exist $\lambda_1(t,x,\xi),\dots,\lambda_m(t,x,\xi)$, classical symbols of order 1 in (x,ξ), C^∞ in t, such that

$$P = (D_t - \lambda_1(t,x,D_x))\dots(D_t - \lambda_m(t,x,D_x)) + R,$$

where $R(x,t,D_x,D_t) = \sum_{j=0}^{m-1} R_j(t,x,D_x)D_t^j$ and $R_j \in S^{-\infty}$.

a) Denote

$$q_j(x,t,\xi,\tau) = \prod_{k\neq j}(\tau - \tau_j(t,x,\xi)).$$

Show that, if

$$a(x,t,\xi,\tau) = \sum_{j=0}^{m-1} a_j(t,x,\xi)\tau^j$$

is homogeneous of degree ℓ in (ξ,τ), one can write

$$a = \sum_{j=1}^{m} b_j(t,x,\xi)q_j(x,t,\xi,\tau),$$

where the b_j are homogeneous of degree $\ell - m + 1$ in ξ. (Give an explicit formula for the b_j.)

b) Verify the statement at the start of Section 2, that

$$P = (D_t - \tau_1(t,x,D_x))\dots(D_y - \tau_m(t,x,D_x)) + R_{m-1},$$

where $R_{m-1}(x,t,D_x,D_t) = \sum_{j=0}^{m-1} R_{m-1,j}(t,x,D_x)D_t^j$ and $R_{m-1,j}$ is classical of order $m-1-j$. Show that there exist classical symbols $\sigma_{1,0}(t,x,\xi),\dots,$ $\sigma_{m,0}(t,x,\xi) \in S^0$, such that

$$\begin{aligned} P = (D_t - \tau_1(t,x,D_x) - \sigma_{1,0}(t,x,D_x)) \quad \dots \quad & (D_t - \tau_m(t,x,D_x) \\ & -\sigma_{m,0}(t,x,D_x))R_{m-2}, \end{aligned}$$

where $R_{m-2}(x,t,D_x,D_t) = \sum_{j=0}^{m-1} R_{m-2,j}(t,x,D_x)D_t^j$ and $R_{m-2,j}$ is classical of order $m-2-j$. Conclude.

Exercise C.5. Let $A = A(t,x,\xi) \in C^\infty([0,T],S^1(\mathbb{R}^n \times \mathbb{R}^n))$ be a symbol with values in the complex square matrices of order m, which depends in a C^∞ way on t.

We are interested in the Cauchy problem

$$(1) \qquad Lu \equiv (D_t - A(t,x,D_x))u = f, \quad u(0) = \varphi \in H^s(\mathbb{R}^n),$$

where all the functions have values in $\mathbb{C}^m$, under the hypothesis (so-called strict hyperbolicity) that all the eigenvalues of the principal symbol $A_1(t,x,\xi)$ of A are real and distinct.

a) Show that there exists $P = P(t,x,\xi) \in C^\infty([0,T], S^0(\mathbb{R}^n \times \mathbb{R}^n))$ with values in the square matrices of order m, which is elliptic of order 0 and such that

$$P(t,x,D_x)A(t,x,D_x)u = A(t,x,D_x)P(t,x,D_x)u + R(t,x,D_x)u,$$

where

$$\Lambda(t,x,\xi) = \begin{pmatrix} \lambda_1(t,x,\xi) & & \\ & \ddots & \\ & & \lambda_m(t,x,\xi) \end{pmatrix}$$

where the principal symbols of the λ_j are the eigenvalues of $A_1(t,x,\xi)$ and where $R \in C^\infty([0,T], S^0(\mathbb{R}^n \times \mathbb{R}^n))$.

b) Prove that there exists a constant C such that, for λ sufficiently large,

$$\sup_{0\le t\le T} e^{-\lambda t}|Pu(t,\cdot)|_s \le C\left(|\varphi|_s \int_0^T e^{-\lambda t}(|Lu(t,\cdot)|_s + |u(t,\cdot)|_s)\mathrm{d}t\right).$$

Deduce the energy inequality

$$\sup_{0\le t\le T} e^{-\lambda t}|u(t,\cdot)|_s \le C\left(|\varphi|_s \int_0^T e^{-\lambda t}|Lu(t,\cdot)|_s\mathrm{d}t\right)$$

(with a different constant C). We remark, as at the end of the proof of Proposition 2.1, that

$$\sup_{0\le t\le T} e^{-\lambda t}|u(t,\cdot)|_{s-1}$$
$$\le |\varphi|_{s-1} + 2\int_0^T e^{-\lambda t}(|Lu(t,\cdot)|_{s-1} + C'|u(t)|_s)\mathrm{d}t.$$

Arguing as in Section 1.2, deduce the existence and uniqueness for the Cauchy problem (1).

Exercise C.6. This problem shows how to reduce a scalar hyperbolic problem of order m to a system of the type considered in Exercise C.5 (the method is due to A.P. Calderón).

Let

$$P(x,t,D_x,D_t) = \sum_{j=0}^{m} P_j(t,x,D_x)D_t^j$$

be a scalar differential operator of order m with $P_m(t,x,\xi) = 1$.

a) We suppose that $Pu = f$, $D_t^{j-1}u(0,x) = \varphi_j(x)$ for $j = 1,\dots,m$. Show that $v = (v_1,\dots,v_m)$ defined by

$$v_j = D_t^{j-1}\langle D\rangle^{m-j}u$$

(where $\langle D\rangle$ has symbol $(1+|\xi|^2)^{1/2}$) is a solution of a system of type

$$(D_t - A(t,x,D_x))v = g, \quad v(0,x) = \psi(x),$$

which should be expressed explicitly.

b) Find the characteristic polynomial of the principal symbol $A_1(t,x,\xi)$ of $A(t,x,D_x)$, and deduce that, if P is strictly hyperbolic, then so too is the system $D_t - A$ in the sense of Exercise C.5. Deduce a study of the Cauchy problem for P, using the results of Exercise C.5

Exercise C.7 (*). (Positive symmetric systems on the torus) This exercise is essentially borrowed from J. Moser [Mo]. On the torus $\mathbb{T}^n$, we consider an $m \times m$ differential system

$$L = \sum_{j=1}^{n} A_j(x)\partial_j + B(x),$$

and we suppose that the A_j are Hermitian. We denote

$$K(x) = \frac{B(x) + B(x)^*}{2} - \frac{1}{2}\sum_{j=-1}^{n} \partial_j A_j(x),$$

and we make the following positivity hypothesis:

$$(P_0) \qquad \forall x \in \mathbb{T}^n, K(x) \geq c_0 I,\ c_0 > 0.$$

a) Show that, for any function $u \in H^1(\mathbb{T}^n)$, we have

$$(1) \qquad \operatorname{Re}(Lu,u) = (Ku,u).$$

Show that this identity remains true if we assume only $u \in L^2$, $Lu \in L^2$. (Use Friedrich's Lemma, Exercise I.5.10 b).)

Deduce the estimate $|u|_0 \leq \text{Const.}|Lu|_0$, then the fact that $u \in L^2$ and $Lu = 0$ implies $u = 0$.

b) Derive the inequality $|u|_0 \leq \text{Const.}|L^*u|_0$, and deduce that, for all $f \in L^2$, the equation $Lu = f$ has precisely one solution $u \in L^2$.

c) Let $\langle D\rangle^s$ denote the operator with symbol $(1+|\xi|^2)^{s/2}$ $(s > 0)$. Assume that the following condition is satisfied:

$$(P_s) \qquad \forall \xi \in S^{n-1}, K(x) + s \sum_{1\leq j,k\leq n} \partial_k A_j(x)\xi_j\xi_k \geq c_s I,$$

where $c_s > 0$, and determine the estimate

$$(2) \qquad \forall u \in H^{s+1}, \operatorname{Re}(\langle D\rangle^s Lu, \langle D\rangle^s u) \geq \frac{c_s}{2}|u|_s^2 - C|u|_0^2$$

(calculate the principal symbol of $[L, \langle D\rangle^s]$, and use Gårding's inequality (Proposition I.5.3)) and deduce that $|u|_s \leq \text{Const.}\,|Lu|_s$.

d) Still under the hypothesis (P_s), we suppose that $f \in H^s$ and consider the solution $u \in L^2$ of $Lu = f$. It is proposed to show that $u \in H^s$.

i) For $\varepsilon > 0$, denote $L_\varepsilon = L - \varepsilon\Delta \mathrm{id}$ ($\Delta = \sum_{j=1}^n \partial_j^2$). Arguing as in a) and b), show that there exists a unique $u_\varepsilon \in L^2$ such that $L_\varepsilon u_\varepsilon = f$. Show that (u_ε) is bounded in L^2 and that a subsequence of (u_ε) tends weakly to u if ε tends to 0.

ii) Show that $u_\varepsilon \in H^{s+2}$ for all $\varepsilon > 0$ and, using the estimate (2), show that (u_ε) is bounded in H^s when ε tends to 0. Conclude.

e) Verify that the condition (P_s) cannot hold for all s if the A_j are constants.

f) Question e) allows one to foresee a limitation on the H^s regularity of the solution of $Lu = f$ for $f \in C^\infty$. Calculate explicitly the solution $u \in L^2$ of

$$\left(-\sin x \frac{\mathrm{d}}{\mathrm{d}x} + m\right) u = (\sin x)^m$$

on $\mathbb{T}^1$, with $m > 1/2$, and verify that the limitation foreseen by the above theory is optimal.

Chapter III

Implicit function theorems

Here we present various examples of perturbation problems, grouped according to the severity of the 'loss of derivatives' incurred: first we have the 'elliptic' cases, where the usual implicit function theorem is sufficient; then there are the problems which are solvable using a fixed-point method (including the question of isometric embedding); finally we have the case in which 'tame estimates' on the linearized form are available to us, which enable us to put in place a Nash–Moser technique (as in the study of perturbations of vector fields on the torus, for example).

A. Implicit function theorem and elliptic problems

1. Review of the implicit function theorem for Banach spaces.

1.1. *Local inversion theorem.*

Theorem. *Let* $f : B_1 \supset U \to B_2$, *where* B_i *is a Banach space and* U *an open subset of* B_1, *and suppose that* f *is continuously differentiable on* U. *If there exists* $A : B_2 \to B_1$ *which is linear and continuous with* $f'(x_0)A = \mathrm{id}_{B_2}$, *then there exists* g, *which is* C^1 *in the neighbourhood of* $y_0 = f(x_0)$, *such that* $f(g(y)) = y$. *If, moreover,* $f'(x_0)$ *is an isomorphism from* B_1 *to* B_2, *then* f *is a* C^1 *diffeomorphism from a neighbourhood of* x_0 *onto a neighbourhood of* $f(x_0)$.

Proof. a) Let us first suppose that $B_1 = B_2$, $f(x_0) = x_0$, $f'(x_0) = \mathrm{id}$. We then write the equation $y = f(x)$ in the form $x = x + y - f(x) \equiv h(x)$, and we look for a fixed point of h as the limit of the sequence $x_{n+1} = h(x_n)$. More precisely,

i) for $\delta > 0$ sufficiently small and $\|y - x_0\| \leq \delta/2$, $h : \overline{B(x_0, \delta)} \to \overline{B(x_0, \delta)}$; indeed

$$f(x) = f(x_0) + x - x_0 + o(\|x - x_0\|) = x + o(\|x - x_0\|),$$

where

$$h(x) - x_0 = y - x_0 + o(\|x - x_0\|), \ \|h(x) - x_0\| \leq \delta(1/2 + \varepsilon(\delta)) \leq \delta;$$

ii) if $\|y - x_0\| \leq \delta/2$ and δ is sufficiently small, then h is a contraction in $\overline{B(x_0, \delta)}$, since

$$\|h(x_1) - h(x_2)\| \leq \|x_1 - x_2\| \sup_{\|z - x_0\| \leq \delta} \|\mathrm{id} - f'(z)\| \leq \frac{1}{2}\|x_1 - x_2\|.$$

Following i) and ii), h has a unique fixed point in $\overline{B(x_0, \delta)}$. Consequently, for any $y \in B(x_0, \delta/2)$, the equation $f(x) = y$ has precisely one solution in $\overline{B(x_0, \delta)}$. Since δ can be chosen arbitrarily small, we deduce that the image under f of any neighbourhood of x_0 is a neighbourhood of x_0.

b) Let us now pass to the general case. Applying the above to $\tilde{f}(y) = f(x_0 + A(y - y_0))$, we first obtain that the image under f of any neighbourhood of x_0 is a neighbourhood of y_0. Let Y be an open ball with centre y_0 and sufficiently small radius, such that

$$\sup_{y \in Y} \|\mathrm{id} - \tilde{f}'(y)\| < 1.$$

Then, for any $y \in Y$, $\tilde{f}'(y)$ is invertible, whence, applying the above to $y \in \tilde{f}$, we conclude that the image under $\tilde{f}$ of any open set in Y is an open set. Since $\tilde{f}$ is also an injection on Y by virtue of the Taylor's formula already used in a), it provides a C^1-diffeomorphism from Y to $\tilde{f}(Y)$. If $\tilde{g}$ is the diffeomorphism, the formula $g(y) = x_0 + A(\tilde{g}(y) - y_0)$ provides the desired mapping g.

c) If $f'(x_0)$ is an isomorphism from B_1 to B_2, then A is invertible, and, following b),

$$f(x) = \tilde{f}(y_0 + A^{-1}(x - x_0))$$

defines a C^1-diffeomorphism from a neighbourhood of x_0 to a neighbourhood of y_0. □

1.2. *Implicit function theorem.* This is a standard corollary of the local inversion theorem.

Theorem. *Let B_0, B_1, B_2 be three Banach spaces, U a neighbourhood of $(x_0, y_0) \in B_0 \times B_0 \times B_1$, and $f : U \to B_2$ continuously differentiable. Suppose that there exists a continuous linear mapping $A : B_2 \to B_1$, such that $f'_y(x_0, y_0)A = \mathrm{id}_{B_2}$. Then there exists $g \in C^1$, from a neighbourhood of x_0 in B_1, such that $f(x, g(x)) = f(x_0, y_0)$, If, in addition, f'_y is bijective, then g is unique and $f(x,y) = f(x_0, y_0)$ is equivalent to $y = g(x)$ for (x,y) near to (x_0, y_0).*

Proof. This is the local inversion theorem applied to $F(x,y) = (x, f(x,y))$. □

This theorem which, as you can see, is easy to prove, turns out to be very powerful for solving certain differential equations: with the help of examples, we shall attempt to understand which.

2. Examples of nonlinear differential equations.

2.1. *Local solvability of differential equations.*

Proposition 2.1. *Let $y' = g(x,y)$ be a system of differential equations, where g is of class C^1 and real in the neighbourhood of $(x_0, y_0) \in \mathbb{R} \times \mathbb{R}^n$. There exists a unique solution y defined on the neighbourhood of x_0, satisfying $y(x_0) = y_0$.*

Proofs. We shall give three proofs, each containing elements which will be useful in more general situations. We can always assume that $x_0 = y_0 = 0$.

First proof. We set $x = \varepsilon t$. If $z(t)$ is a solution of $z' = \varepsilon g(\varepsilon t, z)$ for $t \in [-1, +1]$, with $z(0) = 0$, then $y(x) = z(x/\varepsilon)$ is the desired solution, defined in $[-\varepsilon, +\varepsilon]$.

Suppose $f(\varepsilon, z) = z' - \varepsilon g(\varepsilon t, z)$, considered as a mapping of class C^1 from a neighbourhood of (0,0) in $\mathbb{R} \times B_1$ with values in $B_2 = C^0([-1, +1])$, where $B_1 = \{z \in C^1([-1,+1]), z(0) = 0\}$. Since $f'_z(0,0) = \frac{\mathrm{d}}{\mathrm{d}t}$, we take $A : B_2 \to B_1$ defined by $(Au)(t) = \int_0^t u(s)\mathrm{d}s$, to apply the implicit function theorem.

Second proof. Let $\tilde{y}(x) = xg(0,0)$ be an approximate solution and $I = [-\alpha, +\alpha]$, where $\alpha > 0$ is chosen sufficiently small that $f(y) = y' - g(x,y)$ is well defined in the neighbourhood of $\tilde{y}$ in $B_1 = \{y \in C^1(I), y(0) = 0\}$, with values in $B_0 = C^0(I)$. Since $f'(y) = \frac{\mathrm{d}}{\mathrm{d}x} - g'_y(x, \tilde{y})$ is a system of linear equations, we can find A satisfying $f'(\tilde{y})A = \mathrm{id}_{B_0}$, and the local inversion theorem applies.

Choose $\chi \in C_0^\infty(\mathbb{R})$, with χ equal to 1 near 0: for $\varepsilon > 0$ small, $\chi(x/\varepsilon)f(\tilde{y})$ is a 'small perturbation' in B_0, and we can find $y \in B_1$, near $\tilde{y}$, such that $f(y) = f(\tilde{y}) - \chi(x/\varepsilon)f(\tilde{y})$. In particular, y is the desired solution in a neighbourhood of 0.

Although less stunning than the first proof, this proof demonstrates a procedure (use of an approximate solution at x_0) for transforming a question of local existence at x_0 to a perturbation problem in B_0.

Third proof. We solve the equation

$$y(x) = \int_0^x g(s, y(s))\mathrm{d}s \equiv H(y)(x)$$

seeking a fixed point of H in the ball $\overline{B(0,\beta)}$ of $C^0([-\alpha,\alpha])$. In fact, if $\beta \geq \sup|g|$, H operates in this ball, and if $\alpha \sup|g'_y| < 1$, H is a contraction.

This fixed-point method will be useful in solving hyperbolic systems in Section B.1. □

2.2. *Periodic solutions of nonlinear equations.*

Proposition 2.2. *Let $p :]a,b[\to]c,d[$ be a strictly increasing C^1 function. For any function $g \in C^0_{2\pi}$ (continuous with period 2π) with values in $]c,d[$, there exists a function $y \in C^1_{2\pi}$, with values in $]a,b[$, satisfying $y' + p(y) = g$.*

Proof. Let $U \subset C^1_{2\pi}$ be the open set of functions with values in $]a,b[$, and suppose $f : U \to C^0_{2\pi}$ is defined by $f(y) = y' + p(y)$.

a) For all $y \in U$, $f'(y) : C^1_{2\pi} \to C^0_{2\pi}$ (defined by $f'(y) = \frac{\mathrm{d}}{\mathrm{d}y} + p'(y)$) is invertible, since $\int_0^{2\pi} p'(y)\mathrm{d}x \neq 0$. We can see this by explicit calculation of a periodic solution using the method of variation of constants (cf. also Proposition 3.3.1 later). The local inversion theorem then implies that $f(U)$ is an open set in $C^0_{2\pi}$.

b) Let V be the convex open set of $C^0_{2\pi}$ functions with values in $]c,d[$. Since $f(U) \cap V \neq \emptyset$, it suffices to show that $f(U) \cap V$ is closed in V to obtain $f(U) \supset V$.

Thus, let $g \in V$ and $g_j \in f(U)$, $g_j \to g$, $f(y_j) = g_j$; the equation $y'_j + p(y_i) = g_j$ shows that the values $p(\sup_x y_j)$ and $p(\inf_x y_j)$ are values taken by g_j. Since $g_j([0,2\pi])$ is near $g([0,2\pi]) \subset]c,d[$ by hypothesis, we deduce that the values of y_j are contained in a fixed compact subset of $]a,b[$. The equation then implies that $\|y'_j\| \leq$ Const., and we can extract from the sequence of the y_j a subsequence converging to y in C^0; y is periodic, and the equation shows that $y \in C^1_{2\pi}$, with $f(y) = g$. □

Here we see that an ingenious use of the local inversion theorem, com bined with information of a topological nature, can lead to results of global analysis.

2.3. *Perturbation of a nonlinear Dirichlet problem.*

Proposition 2.3. *For a convex compact $A \subset \mathbb{R}^4$, let us define $B_0(A)$ to be the space of continuous functions on A, possessing continuous partial* *derivatives with respect to the variables* x_2, x_3, x_4. *Let* $u \in C^2(I, \mathbb{R})$ *(*$I = [a,b]$*) be a solution of the Dirichlet problem*

$$F(x, u(x), u'(x), u''(x)) = 0, \ u(a) = u(b) = 0,$$

for some $F \in B_0(A)$, *where* A *is a compact neighbourhood of the values of* (x, u, u', u''). *Suppose that* $F'_{u''} > 0$, $F'_u < 0$. *Then for all* $G \in B_0(A)$ *near* F, *there exists a solution* v *of the problem* $G(x, v, v', v'') = 0, v(a) = v(b) = 0$.

Proof. Consider $f(G, v)$ defined in the neighbourhood of (F, u) by

$$f(G,v)(x) = G(x, v(x), v'(x), v''(x))$$

as a mapping from an open subset of $B_0(A) \times C_0^2(I)$ to $C^0(I)$, where $C_0^2(I) = \{u \in C^2(I), u(a) = u(b) = 0\}$. The differential $f'_v(F, u)$ is

$$f'_v(F, u) = F'_{u''}\frac{\mathrm{d}^2}{\mathrm{d}x^2} + F'_{u'}\frac{\mathrm{d}}{\mathrm{d}x} + F'_u.$$

If $f'_v(F,u)w = 0$, w cannot have a strictly positive maximum or a strictly negative minimum because of the hypothesis $F'_{u''} > 0$, $F'_u < 0$, hence $w = 0$.

The general theory of the linear Dirichlet problem then tells us that f'_v is surjective (it is the 'Fredholm alternative'). Let us give an elementary proof of this: let y_1 and y_2 be solutions of $f'_v y_j = 0$ with $y_1(a) = 0, y'_1(a) = 1, y_2(b) = 0, y'_2(b) = 1$. The injectivity of f'_v in $C_0^2(I)$ implies precisely that y_1 and y_2 are independent. We can then find a solution of $f'_v y = h$ in the form $y = \alpha y_1 + \beta y_2$ by the method of variation of constants, in which we can also choose $\alpha(b) = 0$ and $\beta(a) = 0$, that is $y \in C_0^2(I)$.

Application of the implicit function theorem completes the proof. □

2.4. *Local solvability of nonlinear elliptic equations.*

Proposition 2.4. *Let* $F = F(x, u^{(\alpha)})$ *(*$|\alpha| \leq 2$*) be a* C^∞ *function of its* $n + N$ *real arguments (*$N = (n+1)(n+2)/2$*). Suppose* $p_0 \in \mathbb{R}^N$ *is given such that*

i) $F(0, p_0) = 0$,

ii) *the operator* $P = \sum_\alpha \frac{\partial F}{\partial u^{(\alpha)}}(0, p_0)\partial^\alpha$ *is elliptic, that is to say*

$$p_2(\xi) = \sum_{|\alpha|=2} \frac{\partial F}{\partial u^{(\alpha)}}(0, p_0)\xi^\alpha \neq 0$$

for $\xi \in \mathbb{R}^n$, $\xi \neq 0$.

Then in the neighbourhood of $x = 0$, *there exists a real-valued* C^∞ *solution* u *of the equation*

$$F(x, \partial^\alpha u) = 0$$

with

$$(\partial^\alpha u(0))_{|\alpha|\le 2} = p_0.$$

Proof. Suppose $\tilde u \in C^\infty(\mathbb{R}^n)$ with $\partial^\alpha \tilde u(0)_{|\alpha|\le 2} = p_0$. Fix $\rho > 2$, $\rho \notin \mathbb{N}$, and denote

$$\begin{aligned} B_1 &= \{z \in C^\rho(B), \partial^\alpha z(0) = 0, |\alpha| \le 2\}, \\ B_2 &= \{v \in C^{\rho-2}(B), v(0) = 0\}, \end{aligned}$$

with $B = \{x \in \mathbb{R}^n, |x| \le 1\}$.

For $\varepsilon \in \mathbb{R}$ close to 0, $z \in B_1$, set

$$\begin{aligned} f(\varepsilon, z)(y) &= F(\varepsilon y, \tilde u(\varepsilon y) + \varepsilon^2 z(y), (\nabla \tilde u)(\varepsilon y) \\ &\quad + \varepsilon(\nabla z)(y), (\nabla^2 \tilde u)(\varepsilon y) + \nabla^2 z(y)). \end{aligned}$$

Thus we define a mapping f from a neighbourhood of (0,0) in $\mathbb{R} \times B_1$ to B_2, such that $f(0,0) = 0$. The partial derivative $f'_z(0,0)$ is $P_2 = \sum_{|\alpha|=2} \frac{\partial F}{\partial u^{(\alpha)}}(0, p_0)\partial^\alpha$, the principal part of P. Expanding and rotating the axes if necessary, we can always assume (up to sign) that $p_2 = -|\xi|^2$ and $P_2 = \Delta$. To apply the implicit function theorem, we have to find $A : B_2 \to B_1$ such that $\Delta A = \mathrm{id}$. Using an elementary solution of Δ, it is easy to prove that there exists $A_0 : C^{\rho-2}(B) \to C^\rho(B)$ such that $\Delta A_0 = \mathrm{id}$ (see Exercise A.2).

It then suffices to set, for $v \in B_2$,

$$Av(y) = A_0 v(y) - A_0 v(0) - (A_0 v)'(0) \cdot y - (A_0 v)''(0) \cdot \frac{y^2}{2},$$

since

$$\Delta\left((A_0 v)''(0)\frac{y^2}{2}\right) = \mathrm{trace}\,(A_0 v)''(0) = (\Delta A_0 v)(0) = v(0) = 0.$$

We thus obtain $\varepsilon > 0$ and z satisfying $f(\varepsilon, z) = 0$, and the function $u(x) = \tilde u(x) + \varepsilon^2 z(x/\varepsilon)$ is a C^ρ solution of the problem.

The standard regularity results for elliptic equations then allow us to conclude that indeed $u \in C^\infty$ (see Lemma 2.5). □

Below we present a simple proof of the regularity result used under the additional hypothesis that $\rho > 3$.

Lemma 2.5. *Let Ω be an open subset of $\mathbb{R}^n$ and let $u \in C^\rho(\Omega)$ $(\rho \ge 3)$ be a real function which is a solution of the equation*

$$F(x, \partial^\alpha u) = 0 \quad (|\alpha| \le 2),$$

where F is a C^∞ function of its $n + N$ real arguments. Suppose that the operator $P = \sum_{|\alpha|=2} \frac{\partial F}{\partial u^{(\alpha)}}(x, \partial^\alpha u(x))\partial^\alpha$ is elliptic on Ω. Then $u \in C^\infty(\Omega)$.

Proof. Let $x_0 \in \Omega$. We shall show that $u \in C^{\rho+1}$, near x_0. By varying x_0, we deduce that $u \in C^{\rho+1}$ and then, recursively, $u \in C^\infty(\Omega)$.

Thus let $j \in \{1, \dots, n\}$: we denote $v = \partial_j u \in C^{\rho-1}(\Omega)$. Applying ∂_j to the equation, we have $Pv \in C^{\rho-2}(\Omega)$, which, denoting the operator with constant coefficients equal to the value of the coefficients of P at x_0 by P_0, can again be written as

$$P_0 v + \sum_{|\alpha|=2} a_\alpha \partial^\alpha v \in C^{\rho-2}(\Omega),$$

with $a_\alpha \in C^{\rho-2}(\Omega), a_\alpha(x_0) = 0$.

Let $\varphi \in C_0^\infty(\mathbb{R}^n)$ have value 1 near $\{|x| \leq 1/2\}$ and 0 on $\{|x| \geq 1\}$. For $\varepsilon > 0$, we define

$$\varphi_\varepsilon(x) = \varphi\left(\frac{x - x_0}{\varepsilon}\right), \; v_\varepsilon = \varphi_\varepsilon v, \; a_\alpha^\varepsilon = \varphi_{2\varepsilon} a_\alpha.$$

For ε sufficiently small, $v_\varepsilon \in C_0^{\rho-1}(\mathbb{R}^n)$, $a_\alpha^\varepsilon \in C_0^{\rho-2}(\mathbb{R}^n)$, and

$$P_0 v_\varepsilon = \sum_{|\alpha|=2} a_\alpha^\varepsilon \partial^\alpha v_\varepsilon \in C_0^{\rho-2}(\mathbb{R}^n).$$

We now use the following decomposition introduced in the proof of Proposition II.A.2.1.1 i): $\forall a \in L^\infty(\mathbb{R}^n)$, $\forall b \in L^\infty(\mathbb{R}^n)$, there exist operators A and B such that

$$ab = \sum_q (S_q a) b_q + \sum_p (S_{p+1} b) a_p = Ab + Ba,$$

with

$$\forall \sigma \in \mathbb{R}_+ \setminus \mathbb{N}, \; \|A\|_{C^\sigma \to C^\sigma} \leq \text{Const.} \|a\|_0, \; \|B\|_{C^\sigma \to C^\sigma} \leq \text{Const.} \|b\|_0.$$

Applying this result to $a = a_\alpha^\varepsilon$, $b = \partial^\alpha v_\varepsilon$, we deduce

$$P_0 v_\varepsilon + T_\varepsilon v_\varepsilon \in C^{\rho-2}(\mathbb{R}^{2n}),$$

with

$$\forall \sigma \in \mathbb{R}_+ \setminus \mathbb{N}, \; \|T_\varepsilon\|_{C^{\sigma+2} \to C^\sigma} \leq \text{Const.} \sum_{|\alpha|=2} \|a_\alpha^\varepsilon\|_0 = O(\varepsilon).$$

Letting a parametrix of P_0 act *on the left* on the equation, we obtain

$$v_\varepsilon + K_\varepsilon v_\varepsilon \in C^\rho(\mathbb{R}^n),$$

with

$$\forall \sigma \in \mathbb{R}_+ \setminus \mathbb{N}, \; \|K_\varepsilon\|_{C^{\sigma+2} \to C^{\sigma+2}} = O(\varepsilon).$$

Fixing $\varepsilon > 0$ such that $\|K_\varepsilon\|_{C^{\sigma+2} \to C^{\sigma}+2} < 1$ for $\sigma = \rho - 3$ and $\sigma = \rho - 2$, we deduce that $v_\varepsilon \in C^\rho(\mathbb{R}^n)$, hence $v \in C^\rho(\mathbb{R}^n)$ near x_0, which completes the proof. □

Remarks. a) Lemma 2.5 actually holds for $\rho > 2$ (and even for $\rho = 2$), but the proof is a little more technical.

b) For simplicity, the statement of Proposition 2.4 was deliberately restricted to the case of a scalar equation of order 2. We leave it to readers who are interested to verify that the proof given here can be adapted to the case of elliptic differential systems of arbitrary order.

B. Two examples of the use of the fixed-point method

1. An example from fluid mechanics.

1.1. *Quasilinear symmetric hyperbolic systems.* Let us consider the following Cauchy problem:

$$(*)\qquad \begin{aligned} &\partial_t u + \sum A_j(u)\partial_j u + B(u) = 0,\\ &u|_{t=0} = u_0, \end{aligned}$$

in a band $(t, x) \in [0, T] \times \mathbb{R}^n_x$.

Here $u \in \mathbb{R}^N$, and $A_j(u)$ and $B(u)$ are real $N \times N$ and $N \times 1$ matrices, which are C^∞ in an open $G \subset \mathbb{R}^N$ for which $u_0(x) \in G$. We shall make the following hypothesis.

(H) There exists a real symmetric matrix $S(u)$, which is C^∞ in G, such that
 i) $S(u) \geq S_0 \mathrm{id}$, $S_0 > 0$,
 ii) $S(u)A_j(u)$ is symmetric.

It is then easy to see that the eigenvalues of $\sum A_j(u)\xi_j$ are real for $\xi \in \mathbb{R}^n$, $u \in G$: the system is said to be *hyperbolic.* (Cf. Section II.C.2.)

A system satisfying (H) is said to be *symmetrizable hyperbolic* (the symmetry of the matrices SA_j is essential here (cf. Section II.C.1.1). Finally, the fact that the coefficients A_j depend only on u, and not on ∇u, gives the system a particular nonlinear nature, which we term 'quasilinear'.

Such systems exist in nature. For example, the state of a compressible fluid in $\mathbb{R}^n$ described by its density $\rho > 0$ and its velocity $v = (v_1, \ldots, v_n)$ is subject to the following equations, in the case where the entropy is assumed to be constant (it is of little importance what entropy is here):

$$\begin{aligned} \frac{\partial \rho}{\partial t} + \operatorname{div}(\rho v) &= 0,\\ \frac{\partial(\rho v_i)}{\partial t} + \operatorname{div}(\rho v_i v + p e_i) &= 0, \quad i = 1, \ldots, n, \end{aligned}$$

where $e_i = (0, \ldots, 0, 1, 0, \ldots, 0)$ is the ith basis vector and $p = p(\rho)$ is the pressure, which is an increasing function of ρ, assumed known. In this case,

$N = n + 1$, $B \equiv 0$ and the system can be written in the equivalent form

$$\begin{aligned} \frac{\partial \rho}{\partial t} + \operatorname{div}(\rho v) &= 0, \\ \frac{\partial v_i}{\partial t} + v \cdot \nabla v_i + \frac{c^2}{\rho} \frac{\partial p}{\partial x_i} &= 0 \text{ where } p'_\rho = c^2. \end{aligned}$$

Thus here we have

$$u = (\rho, v), \; A_i(u) = \begin{pmatrix} v_i & 0 & \rho & 0 \\ 0 & v_i & & \\ \vdots & & & \\ c^2/\rho & & & v_i \\ \vdots & & & \\ 0 & & & \end{pmatrix}$$

and

$$S(u) = \begin{pmatrix} c^2 & & & \\ & \rho^2 & & 0 \\ & & \ddots & \\ & 0 & & \rho^2 \end{pmatrix}.$$

However, the coercivity hypothesis (H) i) will only be satisfied if the density ρ remains bounded below by a positive number. This can be achieved, either by solving the system in $[0, T] \times \mathbb{T}^n$ (where $\mathbb{T}^n$ is the n-dimensional torus) or by introducing an appropriate local version of the spaces H^s on $\mathbb{R}^n$ (cf. [Ma1]).

We remark that these equations represent the mathematical translation of the conservation of the mass and of the momentum: the special feature about them is that they are written in divergence form. One such system is a very interesting special case of a quasilinear system known as a conservation law system. Readers may consult [Ma1] for an introduction to this theory. We shall not use this additional property of the example.

1.2. *Short-time existence of a solution.*

Theorem 1.2. *Suppose a system* $(*)$ *satisfies* (H). *If* $u_0 \in H^s$, $s \in \mathbb{N}$, $s > n/2 + 1$, *then there exist* $T > 0$ *and a solution* $u \in L^\infty([0, T]; H^s) \cap \operatorname{Lip}([0, T]; H^{s-1})$ *of* $(*)$.

If we compare $(*)$ with Proposition A.2.1, we observe the following fundamental difference: one cannot use the implicit function theorem (first or second proof) simply because (with the notation of Section A.2.1) the derivative 'lost' by the action of f is not 'regained' by solving the linearized equation f'_z. On the other hand, the third proof can be adapted here as follows: we define a sequence $\{u^n\}_{n\in\mathbb{N}}$ by

$$(**)\qquad S(u^k)\partial_t u^{k+1} + \sum S(u^k)A_j(u^k)\partial_j u^{k+1} + S(u^k)B(u^k) = 0,$$
$$u|_{t=0}^{k+1} = S_{\theta_{k+1}}u_0,$$

where $u^0 = S_{\theta_0}u_0$.

Here S_θ denotes a family of regularizing operators operating in the Sobolev spaces H^s with the properties described in Proposition II.A.1.6, and $\theta_k \uparrow +\infty$ is a sequence to be chosen, In fact, $(**)$ means that at each stage one is solving a linear (symmetric hyperbolic) system which only differs from the 'true' linearized system on u^k in terms of order 0 which are of little importance. The aim of replacing u_0 by $S_{\theta_{k+1}}u_0$ is simply to obtain functions u^k of class C^∞, so that the system considered has C^∞ coefficients, which is always nicer to solve. Here, the 'fixed-point' method consists precisely of writing down the scheme $(**)$.

Thus it remains to prove the existence and convergence of the sequence u^k. This proof will use the following two facts (partially proved in Chapter II):

i) The Cauchy problem for a hyperbolic symmetric system $L = S\partial_t + \sum \tilde{A}_j\partial_j$ can be solved in C^∞, with the 'energy estimate'

$$c_0 \sup e^{-\lambda t}|u(t,\cdot)|_0 \le |u(0,\cdot)|_0 + \int_0^T e^{-\lambda t}|Lu(t,\cdot)|_0 dt,$$

where $S \ge c_0 > 0$ and λ satisfies $2c_0\lambda \ge \|\frac{1}{2}\mathrm{div}\tilde{A} - \frac{\partial S}{\partial t}\|_0$. (See Chapter II, Sections C.1.1 and C.1.2.)

ii) We have available linear estimates of nonlinear functions established in Section II.A.2,

It is convenient to present the proof in two parts: control of u^k in 'large norm' and convergence in 'small norm'.

In fact, we shall prove that $u^{k+1} - u^k$ is a convergent series: the only way of doing this is to subtract the corresponding equations $(**)$ and use the energy estimate i); this assumes that the coefficients $S(u^k)$ and $S(u^k)A_j(u^k)$ are bounded in the Lipschitz norm: it is therefore a bound on u^k in a 'large norm' (stronger than the Lipschitz norm) which we shall establish.

1.3. *Control in 'large norm'.* For the large norm we choose the norm

$$|||u|||_s = |||u|||_{s,T} = \sup_{t\in[0,T]} |u(t,\cdot)|_s$$

(the same s as in the statement of Theorem 1.2).

This norm is well suited to the form of the energy estimates i), and moreover $\|\nabla_x u\|_0 \leq C|||u|||_s$, since $s > n/2+1$ (cf. Proposition II.A.1.4). We make the following induction hypothesis (for a $\delta > 0$ to be chosen):

(H_k) for $0 \leq \ell \leq k$, $||\, u^\ell - S_{\theta_0}u_0|||_{s,T} \leq \delta$,

and we shall show that $(H_k) \Rightarrow (H_{k+1})$ if the parameters δ, θ_0 and T are well chosen (since (H_0) is true by definition of u^0).

a) Let us first ensure that we can solve $(**)$: the values of $u_0(x,t)$ belong to a compact subset K of G, since u_0 is bounded; if θ_0 and $1/\delta$ are sufficiently large, the values of u^k thus remain in a fixed compact subset of G on which $S(u) \geq c_0$.

b) Let us introduce $v^{k+1} = u^{k+1} - S_{\theta_0}u_0$ which satisfies the system

$$\begin{aligned} \partial_t v^{k+1} + \sum A_j(u^k)\partial_j v^{k+1} &= -B(u^k) - \sum A_j(u^k)\partial_j S_{\theta_0}u_0, \\ v^k|_{t=0} &= S_{\theta_{k+1}}u_0 - S_{\theta_0}u_0. \end{aligned}$$

To evaluate $|||v^{k+1}|||_s$, we consider the system satisfied by $\partial_x^\alpha v^{k+1}(|\alpha| \leq s)$:

(1.3.1) $$\begin{aligned} & S(u^k)\partial_t\partial_x^\alpha v^{k-1} + \sum (SA_j)(u^k)\partial_j\partial_x^\alpha v^{k+1} \\ &= -S(u^k)\partial_x^\alpha(B(u^k)) - S(u^k)\partial_x^\alpha \left\{\sum A_j(u^k)\partial_j S_{\theta_0}u_0\right\} \\ &\quad -S(u^k)\sum \left\{\partial_x^\alpha(A_j(u^k)\partial_j v^{k+1}) - A_j(u^k)\partial_j\partial_x^\alpha v^{k_1}\right\} \equiv F, \\ & \partial_x^\alpha v^{k+1}|_{t=0} = \partial_x^\alpha\{S_{\theta_{k+1}}u_0 - S_{\theta_0}u_0\}. \end{aligned}$$

The fact ii) then implies the following lemma.

Lemma. *We have the estimate* $|F(t,\cdot)|_0 \leq C(1+|v^{k+1}(t,\cdot)|_s)$, *where* C *is independent of* T *and of* k.

Proof. The first two terms of F are bounded by a constant. The third can be written (up to the bounded factor $S(u^k)$) as

$$\sum_{\alpha \geq \beta, |\beta| \geq 1} *\partial_x^\beta(A_j(u^k))\partial_x^{\alpha-\beta}\partial_j v^{k+1};$$

we can evaluate the L^2 norm of each term by applying Proposition II.A.2.1.2 to the functions $\partial_\ell(A_j(u^k))$ and $\partial_j v^{k+1}$ (∂_ℓ is a 'detached' derivative of ∂_x^β), for the index $s-1$: we find

$$|\Sigma|_0 \leq C(\|\partial_\ell(A_j(u^k))\|_0 |v^{k+1}|_s + |\partial_\ell A_j(u^k)|_{s-1}\|\partial_j v^{k+1}\|_0).$$

Since $\|u^k\|_0 \leq$ Const.,

$$\begin{aligned} \|\partial_\ell(A_j(u^k))\|_0 = \|A_j'(u^k)\partial_\ell u^k\|_0 &\leq \text{Const. } \|\partial_\ell u^k\|_0 \\ &\leq \text{Const. } |||u^k|||_s \\ &\leq \text{Const.} \end{aligned}$$

and $|\partial_\ell A_j(u_k)|_{s-1} \leq \text{Const.}\,(1 + |||u^k|||_s) \leq \text{Const.}$ by Proposition II.A.2.2. Finally, $\|\partial_j v^{k+1}\|_0 \leq \text{Const.}\,|v^{k+1}|_s$, and the lemma follows. □

It is the 'linear' nature of the estimate of F in this lemma which is at the heart of the proof of the theorem, since it allows us control of $|||v^{k+1}||||_s$ using a procedure of absorption from right to left explained in point c) below.

c) To apply the energy equality i) to (1.3.1), we remark that

$$\partial_t u^k + \sum A_j(u^{k-1})\partial_j u^k + B(u^{k-1}) = 0$$

and (H_k) imply $\|\partial_t u^k\|_0 \leq \text{Const.}$, and thus

$$\|\partial_t(S(u^k)) - 1/2\text{div}\,(SA(u^k))\|_0 \leq \text{Const.}$$

For $\lambda \geq \lambda_0$ sufficiently large, we find

$$\begin{aligned} c_0 \sup e^{-\lambda t}|\partial_x^\alpha v^{k+1}(t,\cdot)|_0 &\leq |S_{\theta_{k+1}}u_0 - S_{\theta_0}u_0|_s + \int_0^T e^{-\lambda t}|F(t,\cdot)|_0 \mathrm{d}t \\ &\leq \delta_1 + CT\sup\left\{e^{-\lambda t}(1 + |v^{k+1}(t,\cdot)|_s)\right\}, \end{aligned}$$

where δ_1 can be chosen arbitrarily small, independently of k, if θ_0 is sufficiently large, since $|S_\theta u_0 - u_0|_s \to 0$ when $\theta \to +\infty$. By adding the inequalities obtained for $|\alpha| \leq s$ and $\lambda = \lambda_0$, we can choose T sufficiently small and θ_0 sufficiently large to obtain (H_{k+1}).

1.4. *Convergence in 'small norm'.* By subtraction, we obtain

$$\begin{aligned} &S(u^k)\partial_t(u^{k+1} - u^k) + \sum(SA_j)(u^k)\partial_j(u^{k+1} - u^k) \\ &= -[S(u^k) - S(u^{k-1})]\partial_t u^k - \sum[(SA_j)(u^k) - (SA_j)(u^{k-1})]\partial_j u^k \\ &\quad - [(SB)(u^k) - (SB)(u^{k-1})], \end{aligned}$$

$$(u^{k+1} - u^k)|_{t=0} = (S_{\theta_{k+1}} - S_{\theta_k})u_0.$$

Once again, the energy inequality i) provides a control of the 'small norm' of $u^{k+1} - u^k$:

$$\begin{aligned} &c_0 \sup e^{-\lambda t}|(u^{k+1} - u^k)(t,\cdot)|_0 \\ &\leq |(S_{\theta_{k+1}} - S_{\theta_k})u_0|_0 + C\int_0^T e^{-\lambda t}|(u^k - u^{k-1})(t,\cdot)|_0 \mathrm{d}t. \end{aligned}$$

Since $\alpha_k \equiv |S_{\theta_{k+1}}u_0 - S_{\theta_k}u_0|_0 \leq \text{Const.}\,\theta_k^{-s-1}(\theta_{k+1} - \theta_k)|u_0|_s$, almost any choice of θ_k (for example, $\theta_k = k$) will make the series of the α_k converge. If T is sufficiently small, we find

$$\sup|(u^{k+1} - u^k)(t,\cdot)|_0 \leq \text{Const. }\sup|(u^k - u^{k-1})(t,\cdot)|_0 + \alpha_k,$$

with $C < 1$ (C is fixed, of course), which implies the convergence of u^k in $L^\infty([0,T]; L^2)$ to u.

1.5. *Regularity of the solution and conclusion.* Since $u^k(t,\cdot)$ is bounded in H^s and convergent in L^2 to $u(t,\cdot)$, we in fact obtain $u \in L^\infty([0,T]; H^s)$ (according to a classical argument: there exists a subsequence extracted from u^k which is weakly convergent in H^s to $v \in H^s$; the limit in the sense of distributions is unique, whence $u = v$). Because of the convexity of the norms $|\ |_s$, we in fact have $|(u^k - u)(t,\cdot)|_s \to 0$ for all $s' < s$: since $s > n/2+1$, we can choose $s' > n/2+1$, which shows that $u \in C([0,T]; C^1)$ is a (classical) solution of $(*)$.

We also deduce that $u \in \mathrm{Lip}([0,T]; H^{s-1})$, using the equation. In fact, one can prove that $u \in C([0,T]; H^s) \cap C^1([0,T]; H^{s-1})$.

2. The isometric embedding problem.

2.1. *General presentation.* Let M be a C^∞ compact manifold and g a Riemannian metric on M, that is to say, in local coordinates $(x_1, \dots, x_n)$, g is a quadratic form $g = \sum g_{ij}(x)\mathrm{d}x_i\mathrm{d}x_j$. We seek, for some N, an injective mapping $u : M \to \mathbb{R}^N$ such that

$$(2.1) \qquad g_{ij} = \frac{\partial u}{\partial x_i} \cdot \frac{\partial u}{\partial x_j} \quad \text{(abbreviated to } g = {}^t u' u'\text{)},$$

that is, such that g corresponds via u to the metric induced on $u(M)$ by $\mathbb{R}^N$ with the usual scalar product, denoted by '$\cdot$' or $(\,,\,)$.

Three fairly elementary facts make it possible to reduce this problem to a *perturbation* problem.

i) By increasing N if necessary, there is scarcely any need to worry about the injectivity of u (see Exercise B.1).

ii) The set of metrics representable in the form (2.1) (for some N) is dense in the set of C^∞ metrics (Exercise B.2).

iii) If $u_j : M \to \mathbb{R}^{N_j}$ $(j = 1, 2)$ is of class C^1 and $u : M \to \mathbb{R}^{N_1+N_2}$ is defined by $u(x) = (t_1 u_1(x), t_2 u_2(x))$, then the corresponding metrics satisfy $g = t_1^2 g_1 + t_2^2 g_2$.

Let us now suppose that we know how to solve (2.1) for g near some g_0: for an arbitrary g, we choose t_0 such that $t_0^2 g_0 < g$, and then we write $g = t_0^2(g_0 + f) + (g - t_0^2 g_0 - t_0^2 f)$; f can be chosen arbitrarily small such that $g_2 = g - t_0^2 g_0 - t_0^2 f$ can be represented in the form (2.1) for $u = u_2$ by virtue of ii); since, by hypothesis, $g_1 = g_0 + f$ can then be represented by some u_1, g can be represented by $(t_0 u_1, u_2)$, by virtue of iii).

To simplify matters, we shall assume (this is not difficult) that (by embedding M in a suitable torus) one can always reduce to the case in which in fact $M = \mathbb{T}^n = \mathbb{R}^n/\mathbb{Z}^n$ is the n-dimensional torus. This allows one to have global coordinates on M and gives a clear meaning to (2.1).

2.2. *Reduction to a fixed-point problem.* In Section B.2.1 we saw how the isometric embedding problem for an arbitrary M could be reduced to the following problem: 'for $M = \mathbb{T}^n$, and g near g_0 (to be chosen), find u satisfying (2.1)'.

We can now state the theorem.

Theorem 2.2. *Let g_0 be the metric induced by a free embedding u_0. For all $\rho > 2$ ($\rho \notin \mathbb{N}$) and any metric g of class C^ρ near g_0, there exists an embedding u of class C^ρ which is a solution of the problem.*

By a 'free embedding', we mean a mapping $u_0 : M \to \mathbb{R}^N$ satisfying

$$\text{(2.2.1)} \qquad \begin{array}{ll} \text{the vectors} & \partial_i u_0 \ (i = 1, \dots, n) \\ \text{and} & \partial^2_{ij} u_0 \ (i \le i, j \le n) \text{ are independent.} \end{array}$$

It is easy to establish that such a u_0 exists (Exercise B.3). To prove the theorem, we set $u = u_0 + v$, $g = g_0 + h$, and the equations $(\partial_i u, \partial_j u) = g_{ij}$ can be written as

$$(\partial_i u_0, \partial_j v) + (\partial_j u_0, \partial_i v) + (\partial_i v, \partial_j v) = h_{ij}. \tag{2.2.2}$$

We then use the following lemma.

Lemma. *If $\Delta = \sum \partial_j^2$ denotes the Laplacian on M, then we have the identity*

$$(1 - \Delta)(\partial_i v, \partial_j v) = \partial_i f_j(v) + \partial_j f_i(v) + r_{ij}(v),$$

where $f_i(v) = -(\Delta v, \partial_i v)$, with $r_{ij}(v)$ a quadratic polynomial (with constant coefficients) in the derivatives $\partial^\alpha v$, $|\alpha| \le 2$.

Proof. We have

$$\begin{aligned} &(1 - \Delta)(\partial_i v, \partial_j v) \\ &= -(\partial_i \Delta v, \partial_j) - (\partial_i v, \partial_j \Delta v) + \text{ quadratic polynomial in } \partial^\alpha v \\ &= -\partial_i(\Delta v, \partial_j v) - \partial_j(\partial_i v, \Delta v) + r_{ij}(v). \end{aligned}$$

□

Thus the equations (2.2.2) can thus be written as

$$\partial_j(\partial_i u_0, v) + \partial_j F_i(v) + \partial_i(\partial_j u_0, v) + \partial_i F_j(v) - 2(\partial^2_{ij} u_0, v) = h_{ij} - R_{ij},$$

where we denote $F_i(v) = (1 - \Delta)^{-1} f_i(v)$, $R_{ij} = (1 - \Delta)^{-1} r_{ij}$, and it suffices to solve the system

$$\begin{aligned} (v, \partial_i u_0) &= -F_i(v), \\ (v, \partial^2_{ij} u_0) &= \frac{R_{ij} - h_{ij}}{2}. \end{aligned} \tag{2.2.3}$$

Letting $M(u_0)w$ denote the unique v linear combination of the $\partial_j u_0$, $\partial^2_{ij} u_0$ satisfying

$$(v, \partial_i u_0) = w_i, \quad (v, \partial^2_{ij} u_0) = w_{ij} \quad \text{(by virtue of (2.2.1))},$$

the system (2.2.3) can finally be written in the 'fixed-point' form

$$v = M(u_0) \begin{pmatrix} -F_i(v) \\ \frac{R_{ij} - h_{ij}}{2} \end{pmatrix} = G(v). \tag{2.2.4}$$

It follows immediately that G maps C^ρ to itself, since $(1-\Delta)^{-1}$, as a pseudo-differential operator of order -2, maps $C^{\rho-2}$ to C^ρ. Moreover, if $R > 0$ is sufficiently small, G is a strict contraction on the ball $B_R = \{v \in C^\rho, \|v\|_\rho \leq R\}$, i.e. $\|Gv\|_\rho \leq C\|v\|_\rho$ for $C < 1$ since f_i and r_{ij} are quadratic in v, whence the existence of a solution of (2.2.4). □

In reality, the reduction presented here has only been discovered recently (see [G]). The isometric embedding problem was first solved by Nash in 1956, using an iteration technique taken up again by Moser in 1961, and subsequently known as the Nash–Moser method. While this technique has ultimately proved to be of no use for solving the problem which led to its introduction, it nevertheless remains a fundamental tool for the study of perturbation problems. We describe it in the final section, together with its applications to the isometric embedding problem and to the problem of perturbations of vector fields on the torus.

C. Nash–Moser theorem

1. General presentation. Let M be a compact C^∞ manifold and Φ a mapping from an open set U of $C^\infty(M, \mathbb{R}^p)$ with values in $C^\infty(M, \mathbb{R}^q)$.

First we recall a number of definitions from differential calculus. In general, given two topological vector spaces E_1 and E_2 on $\mathbb{R}$, and U an open subset of E_1, we say that a continuous mapping $\Phi : U \to E_2$ is of class C^1 if there exists a continuous mapping, which is linear in the second variable,

$$\begin{aligned} \Phi' : U \times E_1 &\to E_2, \\ (u, v) &\mapsto \Phi'(u) \cdot v \end{aligned}$$

such that

$$\forall u \in E, \forall v \in E_1 \; \lim_{t \to 0} \frac{1}{t}(\Phi(u + tv) - \Phi(u)) = \Phi'(u) \cdot v.$$

Mappings of class C^k for $k \geq 2$ are then defined recursively, i.e. Φ is C^k if Φ' is C^{k-1}, and, as always, we denote

$$\Phi^{(k)}(u) \cdot (v_1, \ldots, v_k) = \frac{\partial^k}{\partial t_1 \cdots \partial t_k} \Phi(u + t_1 v_1 + \cdots + t_k v_k)|_{t_1 = \cdots = t_k = 0}.$$

The attentive reader will have remarked that, if one restricts oneself to the case of normed spaces, one does not obtain the classical notion of a C^k mapping, but a rather wider notion. It is not our intention to go into such distinctions; thus we shall limit ourselves to observing that the notion introduced here is both manageable and easily verifiable in all the classical examples of functionals on the space C^∞.

This subject being closed, let us return to our problem: for Φ assumed to be of class C^k, the aim is to solve the equation

$$(*) \qquad \Phi(u) = \Phi(u_0) + f$$

where $u_0 \in C^\infty(M, \mathbb{R}^p)$ is given and f is a 'small' perturbation, also given.

The typical situations in which we shall have recourse to the Nash–Moser method are those in which solution of the linearized equation

$$(**) \qquad \Phi'(u)v = g$$

can only be achieved with a large 'loss of derivatives'. To measure this loss, we shall consider not only a single Banach space as in Sections A and B, but a decreasing family A_s $(0 \leq s < +\infty)$ of such spaces with norms $|\ |_s$, and with intersection $\bigcap A_s = C^\infty$. The most usual examples are $A_s = C^s$ (Hölder spaces) or $A_s = H^s$ (Sobolev spaces) which are defined on M using maps.

Let us suppose we have a solution v of $\Phi'(u)v = g$ (denoted by $v = \psi(u)g$) satisfying, for some fixed u, an estimate of the form

$$|v|_s \leq C|g|_{s+r},$$

for all s (in a finite interval), r being a fixed number. In this case, we say that the equation is solved with a 'loss of r derivatives' (for an arbitrary family of spaces A_s, this expression cannot make sense, but no matter).

In reality, we shall see that it is possible to manipulate only the C^∞ functions during the process of solving $(*)$: thus, this is not about a loss of regularity of the function v, in comparison with that of g, but about a loss of *information* about that regularity.

If we initially assume information about $|u_0|_s$ in an interval with index $s \in [0, s_+]$, the informative interval for u_1 will be $[0, s_+ - r]$, $[0, s_+ - 2r]$ for u_2, etc. (where $u_0, u_1, u_2, \dots, u_n \to u$ are the successive approximations obtained for u), with loss of all information before long.

Furthermore, this 'loss of resolution' r is doubled by a loss r' due to the cost of solving $(**)$ in terms of information about the coefficients of the equation, that is about u. Let us suppose, for example, that $v = \psi(u)g$ satisfies an estimate of the form

$$(***) \qquad |v|_s \leq C(|g|_{s+r} + |g|_r(1 + |u|_{s+r'}))$$

for all s with r and r' fixed. Since the method used is Newton's method (slightly modified) which uses $\Phi'(u_n)$ to calculate u_{n+1}, the solution of one step becomes the coefficient of the next, and the loss r' is added to r to decrease the informative interval.

It is the nature of this double loss of derivatives, relative to the family A_s, which determines the applicability of the Nash–Moser theorem: coarsely speaking, it is sufficient for the losses r and r' to be fixed, in which case (***) is said to be a 'tame' estimate. One should be aware that this tameness depends on the family A_s chosen: faced with a given problem, it is not always possible to know whether such a family exists and if a family is known, there is nothing to prove that another would not lead to smaller losses, or indeed to the application of an ordinary fixed point theorem.

We shall discuss the phenomenon of 'loss of derivatives' for two classical examples (Section C.2), while in Section C.3 we shall see why many nonlinear operations lead to tame estimates in the standard spaces.

Finally, in Section C.4 we shall describe the process for solving (*), which is a Newton scheme combined with a regularization of the functions.

2. Two classical examples.

2.1. *Perturbation of a constant field on the torus.* On the torus $\mathbb{T}^n = \mathbb{R}^n/\mathbb{Z}^n$, we use the coordinates of $\mathbb{R}^n$ as local coordinates. We consider a constant field $\omega \in \mathbb{R}^n$ which we perturb by the field $f : \mathbb{T}^n \to \mathbb{R}^n$, assumed 'small'. We wish to know whether the differential equation $\frac{\mathrm{d}x}{\mathrm{d}t} = \omega + f(x)$ for the integral curves of the field $\omega + f$ is equivalent to $\frac{\mathrm{d}x}{\mathrm{d}t} = \omega$, the equation for the unperturbed field.

More precisely, does there exist a diffeomorphism $U : \mathbb{T}^n \to \mathbb{T}^n$, near to the identity, such that $U'(U^{-1})\omega = \omega + f$? (We shall write $U = \mathrm{id} + u$, with $u : \mathbb{T}^n \to \mathbb{R}^n$ small, it being understood that $u(x)$ will be identified with its image in $\mathbb{T}^n$ via the canonical mapping $\pi : \mathbb{R}^n \to \mathbb{T}^n = \mathbb{R}^n/\mathbb{Z}^n$.)

Already, for $n = 1$ and $f = \text{Const.}$, this is impossible if $f \neq 0$; one has to impose a suitable integral condition upon f. To this end, we introduce n fields $p_1, \ldots, p_n \in C^\infty(\mathbb{T}^n, \mathbb{R}^n)$ such that $\int_{\mathbb{T}^n} p_j \mathrm{d}x$ are independent, and we set, for $t \in \mathbb{R}^n$, $U = \mathrm{id} + u$,

$$\Phi(u,t) = U'(U^{-1})\omega + \sum t_j p_j.$$

The equation (*) which we shall solve is then

$$\Phi(u,t) = \Phi(0,0) + f,$$

and it implies that U conjugates ω to $\omega + f - \sum t_j p_j$.

We shall thus prove the following theorem in Section 4:

Theorem. *There exists a neighbourhood W of 0 in $C^\infty(\mathbb{T}^n, \mathbb{R}^n)$ such that, for any field $f \in W$, there exist $t \in \mathbb{R}^n$ close to 0 and a diffeomorphism of $\mathbb{T}^n$ close to the identity which conjugates $\omega + f - \sum t_j p_j$ to ω.*

Let us now describe the 'loss of derivatives' for this problem.

a) Let us calculate $\Phi'(0,0)(v,t)$: here with $U = \mathrm{id} + v$, we have

$$\begin{aligned}\Phi(v,t) - \Phi(0,0) &= U'(U^{-1})\omega + \sum t_j p_j - \omega \\ &= v'(U^{-1})\omega + \sum t_j p_j \\ &= v'(\omega) + \sum t_j p_j + \text{terms quadratic in } v,\end{aligned}$$

whence $\Phi'(0,0)(v,t) = \omega(v) + \sum t_j p_j$, where $\omega(v)$ denotes the action of the field ω differentiating v ($\omega(v) = \sum \omega_j \frac{\partial v}{\partial x_j}$).

b) To solve the equation with constant coefficients $\omega(v) = h$ on $\mathbb{T}^n$, we expand the first two terms in Fourier series:

$$h(x) = \sum_{k \in \mathbb{Z}^n} h_k e^{2\mathrm{i}\pi\langle k,x\rangle}, \quad v(x) = \sum_{k \in \mathbb{Z}^n} v_k e^{2\mathrm{i}\pi\langle k,x\rangle}.$$

The equation becomes

$$\forall k \in \mathbb{Z}^n, \quad h_k = 2\mathrm{i}\pi\langle k,\omega\rangle v_k.$$

In particular, we require $h_0 = 0$ ($\int_{\mathbb{T}^n} h\mathrm{d}x = 0$); in the present case of the equation $\Phi'(0,0)(v,t) = g \Leftrightarrow \omega(v) = g - \sum t_j p_j = h$, this condition can be written as $\int_{\mathbb{T}^n} g\mathrm{d}x = \sum t_j \int_{\mathbb{T}^n} p_j \mathrm{d}x$ and is satisfied by a unique choice of the t_j. Having chosen the t_j in this way, we assume from then onwards that

$$\forall k \in \mathbb{Z}^n, k \neq 0 \Longrightarrow \langle k,\omega\rangle \neq 0. \tag{2.1.1}$$

Under this hypothesis, there exists a unique choice of v_k given by

$$v_0 = 0, \quad v_k = \frac{1}{2\mathrm{i}\pi}\frac{h_k}{\langle k,\omega\rangle} \quad (k \neq 0).$$

Nevertheless, despite (2.1.1), the numbers $\langle k,\omega\rangle$ may be very small: we call these 'small divisors'.

c) Let us now discuss the hypothesis (2.1.1). It is a classical fact (Exercise C.1) that it is equivalent to the density on $\mathbb{T}^n$, of each curve which is a solution of $\frac{\mathrm{d}x}{\mathrm{d}t} = \omega$.

Moreover, in order to obtain reasonable control over the v_k, it is desirable to reinforce (2.1.1) to

$$\exists \tau > 0, \forall k \in \mathbb{Z}^n, k \neq 0, \quad |\langle k,\omega\rangle| \geq \frac{C}{|k|^\tau}. \tag{2.1.1'}$$

This is the object of the following lemma.

Lemma. *If* $\tau > n-1$, *for almost all* $\omega \in \mathbb{R}^n$, *then there exists* $C = C_\omega$, *such that* (2.1.1′) *is satisfied.*

Proof. Let B be a fixed ball and $B_{k,\varepsilon} = \{\omega \in B, |\langle k,\omega\rangle| \leq \varepsilon |k|^{-\tau}\}$. Since $(B_{k,\varepsilon}) \leq \text{Const.}\varepsilon|k|^{-\tau-1}$, we have

$$\left(\bigcup_{k\neq 0} B_{k,\varepsilon}\right) \leq \text{Const.}\varepsilon \sum |k|^{-\tau-1} \leq \text{Const.}\varepsilon,$$

and hence

$$\left(\bigcap_{\varepsilon>0}\bigcup_{k\neq 0} B_{k,\varepsilon}\right) = 0.$$

□

d) Let us now estimate v; we can write

$$v = h_s * E,$$

where

$$h_s = \sum_{k\neq 0} |k|^s h_k e^{2i\pi\langle k,x\rangle},$$

and

$$E(x) = \sum_{k\neq 0} \frac{1}{2i\pi} e^{2i\pi\langle k,x\rangle} \frac{1}{\langle k,\omega\rangle |k|^s}.$$

If E is integrable, by a direct calculation on $h_s * E$, we then have, in Hölderian norms, $\|v\|_a \leq C\|h_s\|_a$, and, on the other hand $\|h_s\|_a \leq C\|h\|_{a+s}$, since $h_s = (\frac{|D_x|}{2\pi})^s h$ (see Exercise A.2).

Finally, the following lemma (Exercise C.2) indicates the values of s for which the function E is integrable:

Lemma. *If* ω *satisfies* (2.1.1′), *and if* $s > \tau > n-1$, *then*

$$\sum_{k\neq 0} \frac{1}{|\langle k,\omega\rangle||k|^s} < +\infty,$$

and E *is continuous.*

To summarize, we have obtained the following estimate: if $s > \tau$, for all g, there exists a solution (v,t) of the linearized equation $(*)$ at 0, $\phi'(0,0)(v,t) = g$, satisfying

$$\|v\|_a + |t| \leq C\|g\|_{a+s}. \tag{2.1.2}$$

In this example, we see that the 'loss of resolution' may be extremely large and depends on the choice of ω, through the values of the 'small divisors' $\langle k, \omega \rangle$.

2.2. *Example of isometric embedding dealt with using the Nash–Moser technique.* Despite the 'simple' solution of the problem discussed in Section B.2, we have found it instructive to indicate Nash's original method here.

$$(*) \qquad \phi(u) = \phi(u_0) + f,$$

where $\phi(u) = {}^t u' u'$ and u_0 is the free embedding.

The derivative of ϕ with respect to u is

$$\phi'(u)v = {}^t u' v' + {}^t v' u'.$$

To solve the equation $\phi'(u)v = g$ (where g is symmetric), we impose the additional equation ${}^t u' v = 0$, which, by differentiation, gives ${}^t u'' v + {}^t u' v' = 0$; it therefore suffices to choose v satisfying ${}^t u' v = 0$ and ${}^t u'' v = -\frac{g}{2}$: if u is close to u_0, these equations form a linear system of maximal rank, whose solution is unique if one chooses v in the space generated by the (independent) vectors $\partial_j u$, $\partial^2_{jk} u$.

The solution of $\phi'(u)v = g$ is then of the form $v = M(u)g$, where $M(u)$ is a matrix whose coefficients are second-order operators applied to u.

Here, the 'loss of derivatives' $r' = 2$ stems from the presence of known coefficients in the linearized equation; following Proposition A.2.1.1 of Chapter II, we have the estimate

$$(2.2.1) \qquad \|v\|_s \leq C_s(\|g\|_s + \|g\|_0 \|u\|_{s+2}).$$

3. Tame estimates. We first introduce the concept of a tame mapping.

3.1. *Tame mappings.* We recall that a neighbourhood of $u_0 \in C^\infty(M)$ is a neighbourhood in C^μ, for some μ; we shall thus call it a μ-neighbourhood of u_0.

Definition. *Let $u \mapsto \Phi(u)$ be a continuous mapping from an open set U in $C^\infty(M, \mathbb{R}^p)$ to $C^\infty(M, \mathbb{R}^q)$. Then Φ is said to be tame if, for any $u_0 \in U$, there exist a neighbourhood V of u_0, numbers $b \geq 0$ and $r \geq 0$, and constants C_s, such that*

$$\forall u \in V, \forall s \geq b, \quad |\Phi(u)|_s \leq C_s(1 + |u|_{s+r}).$$

We note that the estimate is not valid in all of U, but in V, where the numbers C_s, b and r may depend on u_0 and V.

For simplicity, the definition was only given for mappings on $C^\infty(M)$; it extends in an evident way to more general spaces of the same type [H7,H9]. We leave the following two properties as exercises (Exercises C.3 and C.4).

Property 3.1.1. *The composition of two tame mappings is tame.*

Property 3.1.2. *If Φ is linear and tame, then there exist C_s, b and r such that*

$$\forall u \in C^\infty, s \geq b, \quad |\Phi(u)|_s \leq C_s |u|_{s+r}.$$

The crucial point in the definition of a tame mapping is that it requires a *linear* estimate in the $|u|_{s+r}$ norm, although in general Φ is 'nonlinear'. Moreover, this estimate holds for all s, subject to the sole condition that $u \in V$, which brings into play only a finite number of derivatives of u.

Example. Let F be a C^∞ function of its arguments, and set

$$\Phi(u) = F(x, u, \ldots, u^{(\alpha)}, \ldots), \text{ where } u^{(\alpha)} = \partial_x^\alpha u, |\alpha| \leq m,$$

where u is a real C^∞ function on M.

The mapping Φ is tame, since for all u_0, in a C^m-ball around u_0, from Proposition A.2.2 of Chapter II, we have the estimate ($s \geq 0$) $|\Phi(u)|_s \leq C_s(1 + |u|_{s+m})$ (in C^s norms and in H^s norms).

We shall see that, in fact, many natural nonlinear mappings are tame.

3.2. *Some natural tame mappings.* We have already seen in Propositions A.2.1.1 and A.2.2 of Chapter II that the product uv and the composition $F(u)$ ($F \in C^\infty$) are tame for the two families of Hölder and Sobolev spaces.

Two other cases will be useful in practice (see the remark in Section A.1.5 of Chapter II on Hölder spaces with integer index).

Proposition 3.2.1. *If $\alpha \geq 1$ and $u, v \in C^\alpha$, then $u \circ v \in C^\alpha$ and*

$$\|u \circ v\|_\alpha \leq C(\|u\|_\alpha \|v\|_1^\alpha + \|u\|_1 \|v\|_\alpha + \|u\|_0). \tag{3.2.1}$$

Proof. We have

$$\begin{aligned} \|u \circ v\|_\alpha &\leq \|u\|_0 - \|(u' \circ v)v'\|_{\alpha-1} \\ &\leq C(\|u\|_0 + \|u' \circ v\|_{\alpha-1}\|v'\|_0 + \|u' \circ v\|_0 \|v'\|_{\alpha-1}), \end{aligned}$$

following the inequality for the product of two functions. The last term is bounded by $\|u\|_1 \|v\|_\alpha$. If $\alpha \leq 2$, $\|u' \circ v\|_{\alpha-1} \leq C\|u\|_\alpha \|v\|_1^{\alpha-1}$, whence the proposition holds in this case. If $\alpha > 2$, assuming (3.2.1) is already proven for $\alpha - 1$, we find

$$\begin{aligned} &\|u' \circ v\|_{\alpha-1} \|v'\|_0 \\ &\leq C(\|u'\|_{\alpha-1} \|v\|_1^\alpha + \|u'\|_1 \|v\|_{\alpha-1} \|v\|_1 + \|u'\|_0 \|v\|_1), \end{aligned}$$

and it remains to determine an upper bound for the central term. But, by convexity (Section II.A.1.5)

$$\begin{aligned} \|u\|_2\|v\|_{\alpha-1}\|v\|_1 &\le C\left(\|u\|_1^{\frac{\alpha-2}{\alpha-1}}\|u\|_\alpha^{\frac{1}{\alpha-1}}\right)\left(\|v\|_1^{\frac{1}{\alpha-1}}\|v\|_\alpha^{\frac{\alpha-2}{\alpha-1}}\right)\|v\|_1 \\ &\le C(\|u\|_1\|v\|_\alpha)^{\frac{\alpha-2}{\alpha-1}}(\|u\|_\alpha\|v\|_1^\alpha)^{\frac{1}{\alpha-1}}; \end{aligned}$$

since $a^\lambda b^{1-\lambda} \le C(a+b)$, we obtain (3.2.1). □

Proposition 3.2.2. *Let B_1 and B_2 be convex compact subsets of $\mathbb{R}^n$, and let $u : B_1 \to B_2$, $f : B_2 \to \mathbb{R}^n$ be such that $f(u(x)) = x$ for $x \in B_1$. Then for $\alpha \ge 1$, we have*

$$\|u\|_\alpha \le C\|f\|_\alpha, \tag{3.2.2}$$

where C depends only on α and on a bound for $\|u\|_1$, $\|f\|_1$.

Proof. We have $f'(u(x))u'(x) = \mathrm{id}$, whence $u' = (f'\circ u)^{-1}$: since the mapping $A \mapsto A^{-1}$ (invertible matrix A) is C^∞, we have

$$\|u\|_\alpha \le \|u\|_0 + \|u'\|_{\alpha-1} \le C(1+\|f'\circ u\|_{\alpha-1}) \le C\|f'\circ u\|_{\alpha-1},$$

since in any case $\|f'\|_0 \ge \text{Const.} > 0$. Moreover, applying the definitions

$$\|f'\circ u\|_{\alpha-1} \le C(1+\|f'\|_{\alpha-1}\|u\|_1^{\alpha-1}) \le C\|f\|_\alpha$$

if $\alpha - 1 \le 1$, whence (3.2.2) follows in this case. If $\alpha > 2$, Proposition 3.2.1 gives

$$\|f'\circ u\|_{\alpha-1} \le C(\|f'\|_{\alpha-1}\|u\|_1^{\alpha-1} + \|f'\|_1\|u\|_{\alpha-1} + \|f\|_1),$$

and the middle term is estimated recursively by $\|f\|_2\|f\|_{\alpha-1}$: the same 'convexity' calculation as in Proposition 3.2.1 then gives the estimate. □

3.3. *Solutions of differential equations.* The solution of a system of differential equations, considered as a function of the data and the coefficients, turns out to be a tame application.

We illustrate this remarkable fact (intentionally stated vaguely) by two elementary and precise examples.

3.3.1. *An explicit calculation.*

Proposition 3.3.1. *Let $C^\infty_{2\pi}$ be the space of C^∞ 2π-periodic functions. Let $f \in C^\infty_{2\pi}$, $c = \int_0^{2\pi} f\,\mathrm{d}x \ne 0$, and consider the differential equation $y' + fy = k$, where $k \in C^\infty_{2\pi}$. Then there exists a unique solution of this equation $y \in C^\infty_{2\pi}$, and the mapping $(f,k) \mapsto y$ is tame (for $C^\infty_{2\pi}$ with the C^s norms, $s \ge 0$).*

Proof. We calculate everything: if $F(t) = \int_0^t f(x)\mathrm{d}x$, we set $y = e^{-F}z$, and the equation becomes $z' = ke^F$, giving the general solution $y = Ce^{-F} + e^{-F}\int_0^t e^{F(x)}k(x)\mathrm{d}x$. The condition $y(0) = y(2\pi)$ can be written as

$$C = Ce^{-c} + e^{-c}\int_0^{2\pi} e^F k\mathrm{d}x, \text{ whence } C = \frac{1}{e^c - 1}\int_0^{2\pi} e^F k\mathrm{d}x.$$

The explicit formula obtained for y with this choice of C actually defines a mapping $\tilde{S}$ from the open set $U \subset C^\infty([0, 2\pi])$ where $\int_0^{2\pi} f\mathrm{d}x \neq 0$ to $C^\infty([0, 2\pi])$, which is tame, being a composition of tame mappings: $f \mapsto F, F \mapsto e^{-F}$, etc., following Proposition 3.3.1. The restriction $\tilde{S}|_{C^\infty_{2\pi}}$, which maps $C^\infty_{2\pi}$ to itself, is thus also tame. □

3.3.2. *An a priori estimate.* Unlike in the above example, in which we 'deduced' the tameness of the solution based on an explicit formula, here we shall obtain an estimate directly from the equation itself. This procedure is very important, since it extends to broad classes of partial differential equations: elliptic systems on a compact manifold, boundary value problems of all sorts, etc.

Proposition 3.3.2. *Let A be an $(n \times n)$ matrix which is C^∞ on $[a, b]$. Consider the system $Y' + Ay = k$, where Y takes values in $\mathbb{C}^n$ and satisfies the initial condition $Y(a) = Y_0$. Then for all $n \in \mathbb{N}$, we have the estimate*

$$\|Y\|_{n+1} \leq C(\|k\|_n + (\|k\|_0 + |Y_0|)(1 + \|A\|_n)), \tag{3.3.2}$$

where C depends only on $\|A\|_0$.

Proof. a) Let us first prove the inequality $\|Y\|_0 \leq C(\|k\|_0 + |Y_0|)$. For this, we introduce a 'weight', $e^{-\lambda t}$, as in Section II.C, and set $\|Y\| = \|e^{-\lambda t}Y\|_0$. The equation can be written as

$$Y(t) = Y_0 + \int_a^t [k(s) - A(s)Y(s)]\mathrm{d}s,$$

where

$$\begin{aligned}&|e^{-\lambda t}Y(t)|\\ &\leq e^{-\lambda t}|Y_0| + e^{-\lambda t}\int_a^t |k(s)|\mathrm{d}s + e^{-\lambda t}\int_a^t e^{\lambda s}|A(s)|e^{-\lambda s}|Y(s)|\mathrm{d}s\end{aligned}$$

or

$$\|Y\| \leq C_\lambda|Y_0| + C_\lambda\|k\|_0 + \|A\|_0\|Y\|\sup_t\left(e^{-\lambda t}\int_a^t e^{\lambda s}\mathrm{d}s\right).$$

Since this 'sup' is bounded above by $1/\lambda$, the choice of $\lambda = 2\|A\|_0$ leads to the desired estimate and to (3.3.2) for $n = 0$ using the equation.

b) Let us suppose that (3.3.2) is proved for $n-1$: the equation gives

$$\begin{aligned}\|Y\|_{n+1} &= \|Y\|_0 + \|Y'\|_n \\ &\leq \|k\|_n + C(\|A\|_0\|Y\|_n + \|Y\|_0(1+\|A\|_n)) \\ &\leq \|k\|_n + C\left\{\|k\|_{n-1} + (\|k\|_0 + |Y_0|)(1+\|A\|_{n-1})\right\} \\ &\quad + C\left\{(\|k\|_0 + |Y_0|)(1+\|A\|_n)\right\},\end{aligned}$$

whence (3.3.2) holds for n. □

3.4. *Return to classical examples.* a) In the case of the example of Section C.2.1, we only calculated $\Phi'(0,0)$. Let us indicate how to find $\Phi'(u,s)(v,t)$: consider the curve $U_\theta = \mathrm{id} + u + \theta v$, which has tangent v at the point $\mathrm{id}+u$ corresponding to $\theta = 0$; we need to calculate $\frac{\mathrm{d}}{\mathrm{d}\theta}\left\{\Phi(u+\theta v, s+\theta t)\right\}$ at $\theta = 0$, with

$$\Phi(u+\theta v, s+\theta t) = U'_\theta(U_\theta^{-1})\omega + \sum (s_j + \theta t_j)p_j.$$

We set $U_\theta = U_0 \circ T_\theta$. Then $T_0 = \mathrm{id}$, and $v = \dot{U}_0 = U'_0\dot{T}_0$, where the '.' indicates differentiation with respect to θ. Since $U'_\theta(U^{-1})\omega = U_{\theta *}\omega = U_{0*}T_{\theta *}\omega$, we have

$$\dot{U}_{\theta *}\omega = U_{0*}\dot{T}_{\theta *}\omega,$$

and, on the other hand, taking the derivative at $\theta = 0$ of the identity

$$(T_{\theta *}\omega)(T_\theta(x)) = T'_\theta(x)\cdot\omega,$$

we obtain

$$\dot{T}_{0*}\omega(x) = \dot{T}'_0(x)\cdot\omega$$

since ω is constant. Finally,

$$\Phi'(u,s)(v,t) = U_{0*}\dot{T}'_0\omega + \sum t_j p_j, \text{ with } \dot{T}'_0\omega = \omega(\dot{T}_0) = \omega(U_0'^{-1}v).$$

We then deduce the solution of the equation $\Phi'(u,s)(v,t) = g$ and the estimate

$$\|v\|_a + |t| \leq C_a(\|g\|_{s+a} + \|g\|_0\|u\|_{s+a+1}), \tag{3.4.1}$$

where the number s is the same as in Section C.2.1.

We deduce that, once again, the estimate of the solution (v,t) in terms of the given g and the coefficients (u,s) is tame.

Taking into account this estimate, the theorem stated in Section C.2 is a consequence of the Nash–Moser theorem stated below.

b) Similarly, the solution of the linearized equation $\phi'(u)v = g$ for the isometric embedding problem (see Section C.2.2) leads to a solution v depending on the known g and 'coefficients' u via the tame estimate (linear in

g): if u is in a C^2 neighbourhood of u_0, then for all $s \geq 0$,

$$\|v\|_s \leq C_s(\|g\|_s + \|g\|_0\|u\|_{s+2}).$$

3.5. *Conclusion.* The main hypothesis of the Nash–Moser theorem is precisely the fact that Φ and ψ are tame: this is a reasonable hypothesis, as shown by the general results of Sections C.3.1 to C.3.3, together with the study of the historical example given in Sections C.2 and C.3.4.

4. The Nash–Moser theorem. Here, we work in the space $C^\infty(M)$, M being a compact manifold, where the family of spaces considered is that of the Hölder spaces with the norms $\| \ \|_s$. Let $u_0 \in C^\infty(M, \mathbb{R}^p)$, and let Φ be defined in a μ-neighbourhood of u_0 in $C^\infty(M, \mathbb{R}^p)$ (for some μ) with values in $C^\infty(M, \mathbb{R}^q)$. We make the following two hypotheses:

($\mathcal{H}_1$) Φ is of class C^2 (in the weak sense explained in Section C.1) in a μ-neighbourhood of u_0 and satisfies the tame estimate

$$\begin{aligned}\|\Phi''(u)(v_1, v_2)\|_\alpha \quad &\leq \quad C\left\{\|v_1\|_a\|v_2\|_a(1 + \|u\|_{\alpha+b})\right.\\ &\left.+\|v_1\|_a\|v_2\|_{\alpha+c} + \|v_2\|_a\|v_1\|_{\alpha+c}\right\}\end{aligned}$$

for certain numbers $a, b, c \geq 0$ and all $\alpha \geq 0$.

($\mathcal{H}_2$) For u in a μ'-neighbourhood in $C^\infty(M, \mathbb{R}^p)$, there exists a linear mapping $\psi(u) : C^\infty(M, \mathbb{R}^q) \to C^\infty(M, \mathbb{R}^p)$ such that $\Phi'(u)\psi(u) = \mathrm{Id}$, which satisfies the tame estimate

$$\|\psi(u)g\|_\alpha \leq C\left\{\|g\|_{\alpha+\lambda} + \|g\|_\lambda(1 + \|u\|_{\alpha+d})\right\},$$

for numbers $\lambda, d \geq 0$ and all $\alpha \geq 0$.

We remark that a priori Φ and ψ are only defined on C^∞ functions. We can now state the Nash–Moser theorem.

Theorem. *Suppose Φ satisfies ($\mathcal{H}_1$) and ($\mathcal{H}_2$) and that α is such that $\alpha \geq \mu$, $\alpha \geq \mu'$, $\alpha \geq d$, $\alpha > \lambda + a + c$, $\alpha > a + 1/2\sup(\lambda + b, 2a)$, $\alpha \notin \mathbb{N}$. Then the following hold.*

i) *There exists a $(\alpha + \lambda)$-neighbourhood W of the origin such that for $f \in W$, the equation*

$$\Phi(u) = \Phi(u_0) + f \tag{$*$}$$

has a solution $u = u(f) \in C^\alpha$. Moreover

$$\|u(f) - u_0\|_\alpha \leq C\|f\|_{\alpha+\lambda}.$$

ii) *If there exists $\alpha' > \alpha$, a non-integer, such that $f \in W \cap C^{\alpha'+\lambda}$, then the solution constructed, $u(f)$, belongs to $C^{\alpha'}$, with the corresponding estimate. In particular, if $f \in W \cap C^\infty$, then $u(f) \in C^\infty$.*

Here, the equation $(*)$ means that there exists a sequence $u_j \in C^\infty(M)$ such that, for all $\varepsilon > 0$, $u_j \to u$ in $C^{\alpha-\varepsilon}$ and $\Phi(u_j) \to \Phi(u_0) + f$ in $C^{\alpha+\lambda-\varepsilon}$.

4.1. *Solution scheme.* As we said in Section C.1, this scheme is a *Newton scheme* combined with a regularization of the functions considered.

a) A 'pure' Newton scheme would imply the following.

Suppose that $u_0, \ldots, u_n$ are defined, and let us attempt to define u_{n+1} by $u_{n+1} = u_n + \Delta u_n$. For $k \leq n$, with $\Delta u_k = u_{k+1} - u_k$, we write

$$\phi(u_{k+1}) = \phi(u_k) + \phi'(u_k)\Delta u_k + \varepsilon_k,$$

which defines the 'square error' ε_k (at this point, only $\varepsilon_0, \ldots, \varepsilon_{n-1}$ are actually known).

We shall choose Δu_n satisfying $\phi'(u_n)\Delta u_n = \gamma_n$, where γ_n is to be determined such that ε_n measures the distance between $\phi(u_{n+1})$ and the sought target $\phi(u_0) + f$.

Summing the above equations, we obtain (with $\gamma_k = \phi'(u_k)\Delta u_k$)

$$\phi(u_{n+1}) = \phi(u_0) + \sum_0^n \gamma_k + \sum_0^n \varepsilon_k;$$

γ_n will thus be defined by

$$\sum_0^n \gamma_k + E_n = f \tag{4.1}$$

where $E_n = \sum_0^{n-1} e_k$ is the 'accumulated error', so that

$$\phi(u_{n+1}) = \phi(u_0) + f + \varepsilon_n \text{ and } \gamma_n = f + \phi(u_0) - \phi(u_n).$$

Since we expect to find $\varepsilon \to 0$, formally, this equation gives $(*)$.

b) We have seen that such a scheme is impracticable, since the 'informative interval' for Δu_n shrinks like a shagreen skin, with the consequences which we know about.

We then introduce the approximation operators S_θ which were defined and studied in Section II.A.1.6 and which have the following properties ($\theta \geq 1$, α and β in a bounded interval):

i) $\|S_\theta u\|_\beta \leq C\|u\|_\alpha$, $\beta \leq \alpha$,
ii) $\|S_\theta u\|_\beta \leq C\theta^{\beta-\alpha}\|u\|_\alpha$, $\beta \geq \alpha$,
iii) $\|u - S_\theta u\|_\beta \leq C\theta^{\beta-\alpha}\|u\|_\alpha$, $\beta \leq \alpha$,
iv) $\|\frac{\mathrm{d}}{\mathrm{d}\theta} S_\theta u\|_\beta \leq C\theta^{\beta-\alpha-1}\|u\|_\alpha$ for all β.

We choose a sequence $\theta_n = (\theta_0^{1/\varepsilon} + n)^\varepsilon$ ($\varepsilon > 0$, $\theta_0 \geq 1$ to be chosen), and we decompose the 'perturbation' f into

$$f = S_{\theta_0} f + \sum_{k \geq 0} \Delta_k \dot{f}_k,$$

$$\Delta_k = \theta_{k+1} - \theta_k, \quad \dot{f}_k = \frac{1}{\Delta_k}(S_{\theta_{k+1}} - S_{\theta_k})(f).$$

Let us remark that if we had chosen $\theta_n = C2^n$, we would have obtained the usual Littlewood–Paley decomposition of f (see Section II.A.1). The choice of a smaller step Δ_k, as we shall see (Section C.4.3 b)), permits a finer discussion; but the fundamental property of the block $\dot{f}_k$ remains, namely, for all s,

$$\|\dot{f}_k\|_s = \left\| \frac{1}{\Delta_k} \int_{\theta_k}^{\theta_{k+1}} \frac{\mathrm{d}}{\mathrm{d}\theta} S_\theta f \mathrm{d}\theta \right\|_2 \leq C\theta_k^{s-\alpha-1} \|f\|_\alpha. \tag{4.2}$$

Now, instead of seeking, as in a), to reach the target $\Phi(u_0) + f$ with the first blow u_1, we proceed in small steps, targetting only $\Phi(u_0) + S_{\theta_0} f$ with u_1, then $\Phi(u_0) + S_{\theta_1} f$ with u_2, etc. At each stage, and herein lies the fundamental innovation, instead of solving $\Phi'(u_n)\Delta u_n = *$, we solve $\Phi'(v_n)\Delta u_n = *$, where $v_n = S_{\theta_n} u_n$.

The scheme is thus as follows: again, we assume that $u_0, u_1, \ldots, u_n$ are defined and we seek to define u_{n+1} by $u_{n+1} = u_n + \Delta u_n$, $\Delta u_n = \psi(v_n)\gamma_n$ (this is the solution we have for $\Phi'(v_n)$), where γ_n is to be determined. Similarly, for $k \leq n-1$, we set

$$u_{k+1} - u_k = \Delta u_k = \psi(v_k)\gamma_k.$$

As in a), for $k \leq n$, we write

$$\Phi(u_{k+1}) - \Phi(u_k) = \gamma_k + \varepsilon_k,$$

where, this time,

$$\begin{aligned} \varepsilon_k &= \varepsilon_k' + \varepsilon_k'', \\ \varepsilon_k' &= \left\{\Phi'(u_k) - \Phi'(v_k)\right\} \Delta u_k, \\ \varepsilon_k'' &= \Phi(u_{k+1}) - \Phi(u_k) - \Phi'(u_k)\Delta u_k. \end{aligned}$$

The error ε_k'' is the usual 'square error' of the Newton scheme, while ε_k' is a new 'substitution error' which we have introduced by replacing u_k by v_k in Φ'.

Once again setting $E_n = \sum_0^{n-1} \varepsilon_k$ here, we determine γ_n recursively using the formula

$$\sum_0^n \gamma_k + S_{\theta_n} E_n = S_{\theta_n} f. \tag{4.1'}$$

Let us end this section with a final putting in place of notation: in order that the increment Δ_n should not appear systematically in the estimates of the quantities introduced above, we prefer to set, for all n,

$$\Delta u_n = \Delta_n \dot{u}_n,\ \gamma_n = \Delta_n g_n,\ \varepsilon_n = \Delta_n \varepsilon_n,\ \varepsilon'_n = \Delta_n \varepsilon'_n,\ \varepsilon''_n = \Delta_n e''_n,$$

such that

$$\begin{aligned} e'_n &= \{\Phi'(u_n) - \Phi'(v_n)\}\, \dot{u}_n, \\ e''_n &= \frac{1}{\Delta_n}\left\{\Phi(u_{n+1}) - \Phi(u_n) - \Phi'(u_n)\Delta \dot{u}_n\right\}. \end{aligned}$$

4.2. *Structure of the induction.* It only remains now to prove the existence and the convergence of the sequence u_n under the hypothesis that $\|f\|_\beta$ is sufficiently small ($\beta = \alpha + \lambda$).

For a sufficiently large $\delta > 0$ and number $\tilde{\alpha} > \alpha$, both of which will be fixed later, we make the following induction hypothesis:

(H_n) For $0 \le k \le n$ and $s \in [0, \tilde{\alpha}]$ (the 'informative interval')

$$\|\dot{u}_k\|_s \le \delta \theta_k^{s-\alpha-1}.$$

We shall thus assume (H_n) and in Section C.4.3 we shall prove that (H_n) $\Rightarrow$ (H_{N+1}) if the parameters have been well chosen.

Finally, (H_0) will be dealt with separately. Here are a few preliminary remarks.

a) If (H_n) is true for all n, the sequence u_n converges in $C^{\alpha-\varepsilon}$ (for all $\varepsilon > 0$) to a function $u \in C^\alpha$ by virtue of the following fact:

Lemma 4.2.1. *Let $(u_\theta)_{\theta \ge \theta_0}$ be a family of C^∞ functions such that*

$$\|u_\theta\|_{a_j} \le M\theta^{b_j - 1},\ j = 1, 2,\ \textit{with } b_1 < 0 < b_2,\ a_1 < a_2.$$

Let us define ν by $\nu b_1 + (1-\nu) b_2 = 0$ and $a = \nu a_1 + (1-\nu) a_2$; then if $a \notin \mathbb{N}$,

$$u = \int_{\theta_0}^{+\infty} u_\theta \mathrm{d}\theta \in C^a \textit{ and } \|u\|_a \le CM.$$

The point of this lemma is that the integral which defines u converges normally in $C^{a'}$ for $a' < a$, where a is precisely the index for which $\|u_\theta\| \le CM\theta^{-1}$. (The proof is given in Exercise C.8.)

b) Let us explain the significance of (H_n): we will have $u = u_0 + \sum_{k=0}^{+\infty} \Delta_k \dot{u}$, so that we construct u by the inverse of the procedure used to decompose f. The desired estimate (H_n) is nothing other than the estimate (4.2) which we 'are entitled' to expect for a block of u. Of course, except in very special cases, the $\dot{u}_k$ *will not be* the 'standard blocks' of u, but they will play the same role (for example, in the case of the Littlewood–Paley decomposition, a 'standard block' has its spectrum in a shell, which implies (4.2), but not the inverse).

c) Let us show that if $\alpha \geq \mu$, $\alpha \geq \mu'$, $0 < \delta \leq \delta_0$, δ_0 sufficiently small and θ_0 sufficiently large, then (H_n) implies that u_{n+1} and v_{n+1} (and the interval which joins them) are in a sufficiently small ball around u_0, in which we can use $(\mathcal{H}_1)$ and $(\mathcal{H}_2)$. This follows from the lemma below.

Lemma 4.2.2. *(H_n) implies (denoting $x_+ = \sup(x, 0)$)*

i) $\|u_{n+1} - u_0\|_\alpha \leq C\delta$,

ii) $\|S_{\theta_{n+1}}(u_{n+1} - u_0)\|_a \leq C\delta\theta_{n+1}^{(a-\alpha)_+}$ *for a in a finite interval,*

iii) $\|(u_{n+1} - u_0) - S_{\theta_{n+1}}(u_{n+1} - u_0)\|_a \leq C\delta\theta_{n_1}^{a-\alpha}$, $0 \leq a \leq \tilde{\alpha}$,

iv) $\|u_{n+1} - v_{n+1}\|_a \leq C\theta_{n+1}^{a-\alpha}$, $0 \leq a \leq \tilde{\alpha}$,

v) $\|v_{n+1}\|_a \leq C\theta_{n+1}^{(c-\alpha)_+}$, *for a in a finite interval.*

Proof. Let $U = u_{n+1} - u_0 = \sum_{j=0}^{n} \Delta_j \dot{u}_j$. The lemma implies that $\|U\|_\alpha \leq C\delta$, whence we have i) and iii) when $a \leq \alpha$.

On the other hand, for $a = \tilde{\alpha}$

$$\|U\|_a \leq \delta \sum_{j=0}^{n} \Delta_j \theta_j^{a-\alpha-1} \leq C\delta\theta_{n+1}^{a-\alpha},$$

whence iii) follows.

We write

$$u_{n+1} - v_{n+1} = u_0 + U - S_{\theta_{n+1}}(u_0 + U) = u_0 - S_{\theta_{n+1}}u_0 + U - S_{\theta_{n+1}}U,$$

which proves that iii) implies iv); finally i) implies ii) and

$$v_{n+1} = S_{\theta_{n+1}}(u_0 + U),$$

whence v) follows. □

Since $\alpha \geq \mu$ and $\alpha \geq \mu'$, i) shows that u_{n+1} is in the intersection of the μ and μ' neighbourhoods required by $(\mathcal{H}_1)$ and $(\mathcal{H}_2)$ if $\delta \leq \delta_0$; moreover,

$$v_{n+1} - u_0 = S_{\theta_{n+1}}(u_{n+1} - u_0) + S_{\theta_{n+1}}u_0 - u_0$$

and ii) shows that the same is true for v_{n+1} if in addition θ_0 is sufficiently large

We remark that the condition $\alpha \notin \mathbb{N}$ of the theorem is only used here to obtain Lemma 4.2.1.

4.3. *Proof of the induction hypothesis.* In the previous section, u_{n+1} and v_{n+1} were guaranteed to remain in a fixed neighbourhood of u_0, so that it was possible to define $\dot{u}_{n+1}$ and use $(\mathcal{H}_1)$ and $(\mathcal{H}_2)$; we shall now prove (H_{n+1}) assuming (H_n).

This is fastidious work which first involves evaluating the errors e_n, E_{n+1}, then g_{n+1} and finally $\dot{u}_{n+1}$; in doing this, one has to pay great attention to the 'informative intervals' for the estimates obtained.

a) Estimate for e'_k.

Lemma. *For $0 \leq k \leq n$, $s \in [0, \tilde{\alpha} - \sup(b, c)]$, we have*

$$\|e'_k\|_s \leq C\delta\theta_k^{L(s)-1}, \tag{4.3.1}$$

where $L(s) = \sup(s + a + c - 2\alpha, (s + b - \alpha)_+ + 2a - 2\alpha)$.

Proof. We have

$$e'_k = \left\{\Phi'(u_k) - \Phi'(v_k)\right\}\dot{u}_k = \int_0^1 \Phi''(v_k + t(u_k - v_k))(\dot{u}_k, u_k - v_k)\mathrm{d}t,$$

whence, by virtue of $(\mathcal{H}_1)$,

$$\begin{aligned}\|e'_k\|_s &\leq C\|\dot{u}_k\|_a\|u_k - v_k\|_a(1 + \sup_t \|v_k + t(u_k - v_k)\|_{s+b}) \\ &\quad + C\|\dot{u}_k\|_a\|u_k - v_k\|_a + C\|\dot{u}_k\|_{c+s}\|u_k - v_k\|_a \\ &\leq C\delta\theta_k^{L(s)-1}\end{aligned}$$

where $L(s) = \sup(s + a + c - 2\alpha, (s - b - \alpha)_+ + 2a - 2\alpha)$. □

b) Estimate of e''_k. We have

$$e''_k = \Delta_k \int_0^1 (1 - t)\Phi''(u_k + t\Delta_k\dot{u}_k)(\dot{u}_k, \dot{u}_k)\mathrm{d}t,$$

with an estimate for the integral of the same form as for e'_k on the same interval; it is then clear that a suitable choice of $\varepsilon > 0$ makes e''_k negligible in comparison with e'_k, since $\Delta_k \sim \varepsilon\theta_k^{1-1/\varepsilon}$. The error e_k is thus estimated by (4.3.1).

c) Estimate of g_{n+1}. We recall that

$$g_{n+1} = \frac{\Delta_n}{\Delta_{n+1}}\left\{\dot{f}_n - (\dot{E}_n) - S_{\theta_{n+1}}e_n\right\},$$

where, for any function w, the block $\dot{w}_n = \frac{1}{\Delta_n}(S_{\theta_{n+1}} - S_{\theta_n})w$ is estimated for any index s, by the inequality (4.2). In particular here

$$\|\dot{f}_n\|_s \leq C\theta_n^{2s-\beta-1}\|f\|_\beta \quad (\beta = \alpha + \lambda)$$

and

$$\|(\dot{E}_n)_n\|_s \leq C\theta_n^{s-r-1}\|E_n\|_r;$$

we choose r to be as large as possible, namely $r = \tilde{\alpha} - \sup(b, c)$, and $\tilde{\alpha}$ to be sufficiently large that the series $\sum \Delta_j \theta_j^{L(r)-1}$ diverges (i.e. $L(r) > 0$) and that L has gradient 1 to the right of r. We then have $\|E_n\|_r \leq C\delta\theta_n^{L(r)}$ and

$$\|(\dot{E}_n)_n\|_s \leq C\delta\theta_n^{s-r-1+L(r)} \leq C\delta\theta_n^{L(s)-1}, \tag{4.3.2}$$

since $L(r) \leq L(s)+r-s$ if $s \leq r$ (because $|L'| \leq 1$), while $L(r) = L(s)+r-s$ if $s \geq r$ by the choice of $\tilde{\alpha}$. Similarly, $\|S_{\theta_{n+1}} e_n\| \leq C\delta\theta_n^{L(s)-1}$, for all s. It is precisely with a view to obtaining estimates for g_{n+1} on an arbitrary (bounded) 'index interval' that we substituted $S_{\theta_n} f$ for f and $S_{\theta_n} E_n$ for E_n in (4.1) (whence (4.1′)). Finally, we find

$$\|g_{n+1}\|_s \leq C\left(\delta\theta_{n+1}^{L(s)-1} + \theta_{n+1}^{s-\beta-1}\|f\|_\beta\right),$$

for s in a bounded interval.

d) Estimate for $\dot{u}_{n+1}$. By definition, $\dot{u}_{n+1} = \psi(v_{n+1})g_{n+1}$, whence for $s \in [0, \tilde{\alpha}]$,

$$\|\dot{u}_{n+1}\|_s \leq C\left\{\|g_{n+1}\|_{s+\lambda} + \|g_{n+1}\|_\lambda(1 + \|v_{n+1}\|_{s+d})\right\},$$

or

$$\begin{aligned}\|\dot{u}_{n+1}\|_s \;\leq\; & C\delta\theta_{n+1}^{L(s+\lambda)-1} + C\|f\|_\beta\theta_{n+1}^{s-\alpha-1} \\ & + C(\delta\theta_{n+1}^{L(\lambda)-1} + \|f\|_\beta\theta_{n+1}^{-\alpha-1})\theta_{n+1}^{(s+d-\alpha)_+}.\end{aligned}$$

We remark that the replacement of u_{n+1} by v_{n+1} enables us to avoid the shrinking of the information interval due to the 'cost of information about the coefficients' in the solution of Φ'.

We will obtain (H_{n+1}) (there is no constant in front of δ in (H_{n+1})!) provided that, for $s \in [0, \tilde{\alpha}]$,

i) $L(s + \lambda) < s - \alpha$,
ii) $\|f\|_\beta/\delta$ is small,
iii) $L(\lambda) + (s + d - \alpha)_+ < s - \alpha$,
iv) $(s + d - \alpha)_+ - \alpha \leq s - \alpha$,
v) θ_0 is sufficiently large.

Let us analyse these conditions: in order to satisfy iv), it is necessary to take $\alpha \geq d$, which is sufficient; iii) is then equivalent to $L(\lambda) < -\alpha$, while i) is implied by $L(\Lambda) < -\alpha$. Finally, $L(\lambda) < -\alpha$ implies that

$$\lambda + a - c - 2\alpha < -\alpha \Leftrightarrow \alpha > \lambda + a + c,$$

$$(\lambda + b - \alpha)_+ + 2a - 2\alpha < -\alpha \Leftrightarrow a > a + 1/2\sup(\lambda + b, 2a).$$

The crucial point is the following: the quadratic nature of the errors is reflected in the nature of the estimate (4.3.1) by the presence of the term in

'-2α' in $L(s)$, while the tame nature of the estimates implies $|L'| \leq 1$; it is precisely this '-2α' term which allows us to choose α (sufficiently large) to 'close' the induction.

From now on θ_0 will be fixed taking into account v).

e) How to obtain (H_0). Since $g_0 = \frac{1}{\Delta_0} S_{\theta_0} f$,

$$\begin{aligned} \dot{u}_0 &= \frac{1}{\Delta_0} \psi(S_{\theta_0} u_0) S_{\theta_0} f, \\ \|\dot{u}_0\|_s &\leq \frac{C}{\Delta_0} \{\|S_{\theta_0} f\|_{s+\lambda} + \|S_{\theta_0} f\|_\lambda (1 + \|S_{\theta_0} u_0\|_{s+d})\}; \end{aligned}$$

θ_0 is already fixed, and we obtain (H_0) simply by choosing f/δ sufficiently small in any norm. Taking into account the condition ii) of d), we take $\|f\|_\beta \delta$ sufficiently small, and assertion i) of the theorem is proved. □

4.4. *Additional regularity of the solution constructed.* Let us now prove assertion ii) of the Nash–Moser theorem. Let us suppose that $f \in W \cap C^{\alpha'+\lambda}$ for some $\alpha' > \alpha$ and choose $\tilde{\alpha} > \alpha'$. We shall use the estimates (H_n), which are true for all n, to obtain

$$(H'_n) \qquad \forall s \in [0, \tilde{\alpha}], \quad \|\dot{u}_n\|_s \leq \text{Const.} \theta_n^{s-\alpha'-1}.$$

Lemma 4.2.1 will then allow us to complete the proof as before.

We remark that, in the estimate for $\dot{u}_{n+1}$ deduced from (H_n) (see Section C.4.3 d)), the exponent of θ_{n+1} in the terms in which $\|f\|_\beta$ does not occur is *strictly* less than $s - \alpha - 1$. On the other hand, the presence of $\|f\|_\beta$ in this inequality stems from an estimate for $\dot{f}_n$ (see Section C.4.3 c)); we can therefore replace $\|f\|_\beta \theta_{n+1}^{-\alpha-1}$ by $\|f\|_{\sigma+\lambda} \theta_{n+1}^{-\sigma-1}$, for all $\sigma > 0$. We have thus proved that, for all n, for all $s \in [0, \tilde{\alpha}]$,

$$\|\dot{u}_{n+1}\|_s \leq \text{Const.}\, (\theta_{n+1}^{s-\alpha-\gamma-1} + \|f\|_{\beta'} \theta_{n+1}^{s-\alpha'-1}),$$

with $\gamma > 0$ and $\beta' = \alpha' + \lambda$. In particular (ignoring the trivial case $n = 0$)

$$\forall n, \ \|\dot{u}_n\|_s \leq \text{Const.}\, \theta_n^{s-\rho-1},$$

with $\rho = \min(\alpha + \gamma, \alpha')$. With these new estimates instead of (H_n), we can revisit the same proof, remarking that the gain γ depends only on a lower bound for α and thus is equally valid for all $\rho \geq \alpha$. Thus, after k iterations, we obtain

$$\forall n, \ \forall s \in [0, \tilde{\alpha}], \quad \|\dot{u}_n\|_s \leq \text{Const.}\, \theta_n^{s-\rho_k-1},$$

with $\rho_k = \min(\alpha + k\gamma, \alpha')$.

For k sufficiently large, we obtain $\rho_k = \alpha'$, which gives the desired estimates (H'_n). □

4.5. *Conclusion and remarks.* We can summarize the idea of the proof heuristically as follows: by only allowing regularizations of solutions to appear in the Newton scheme, we avoid the shrinking of the informative interval at the price of poorer estimates; but since these tame estimate are 'linear in large norms', the quadratic nature of the errors of the Newton scheme compensates for this loss.

We have only given the proof for the $C^\infty(M)$ case, taking the Hölder spaces for the family of spaces. The reason for this is that it is nice to have at one's disposal Lemma 4.2.1 which gives the 'proper' estimates of Lemma 4.2.2.

Of course, it is no problem to adapt the procedure described to numerous other situations and other families of spaces: one must however be aware of the fact that Lemma 4.2.1 is then no longer true, which results in a small loss of $\varepsilon > 0$ (arbitrary) in the estimate of Lemma 4.2.2. (A more sophisticated approach involves 'correcting' the family of spaces used; cf. [H7].)

Commentary on Chapter III

This chapter is intended to provide the reader with practical means of detecting the 'Nash–Moser symptom' in a given pathological case.

It contains numerous examples divided into three classes. In Section A, we study 'elliptic' situations, that is to say cases in which the solution of the linearized equation enables us to 'regain' all the lost derivatives. These cases are handled using the conventional implicit function theorem: certain examples are due to Hamilton [Ha].

In Section B, we discuss situations which do not go back to the implicit function theorem, but in which the 'loss of derivatives' is not severe and enables one to put in place a fixed-point scheme. We found the case of quasilinear hyperbolic systems in fluid mechanics typical and non-academic: the presentation given here is inspired by Majda [Ma1]. The handling of the isometric embedding problem by a fixed-point method is a recent contribution due to Günther [G].

Finally, Section C is given over to the Nash–Moser theorem, which allows arbitrary, but fixed, losses. Here again, our debt to Hörmander is large, since Section 4.4 has its origins in his course at Stanford (1977, unpublished) and his article [H6]. We have striven to show how the scheme which he proposes is obtained naturally by modifying the Newton scheme slightly.

Of course, there are numerous other presentations of the theorem which we have not wished to discuss here: readers might consult, for example, Moser [Mo] and will find a detailed discussion of 'tame applications', 'tame mappings', etc. in Hamilton [Ha], together with numerous examples and references.

This chapter is in a sense the outcome of the two previous chapters, which contained a leisurely presentation of the fundamental concepts used (tame estimates, energy inequalities, etc.). It also underlines the fact that the core of a nonlinear problem of 'perturbation' type is the solution (with tame control) of a linear problem, which is often pseudo-differential in nature. Majda [Ma2] contains another remarkable illustration of this fact, which is, unfortunately, too erudite to be presented here.

Exercises for Chapter III

Exercise A.1. Let D be the unit disc in $\mathbb{C}$ and $W(D)$ the space of holomorphic functions f on D satisfying

$$\|f\|_{W(D)} = \sum_{n=0}^{\infty} \frac{|f^{(n)}(0)|}{n!} < +\infty.$$

a) Show that if $f \in W(D)$, and if g is holomorphic near the values taken by f on $\overline{D}$, then $g \circ f \in W(D)$. (Remark that $\|h\|_{W(D)} \leq C \sup_{x \in D}(|h(z)| + |h''(z)|)$ and that $W(D)$ is an algebra; then write

$$f = f_1 + f_2, \text{ with } f_2(z) = \sum_{n>N} \frac{f^{(n)}(0)}{n} z^n,$$

where N is chosen sufficiently large that the series $\sum_{n=0}^{\infty} \frac{g^{(n)}(f_1)}{n!} f_2^n$ is well defined and converges in $W(D)$.)

b) Let f be a holomorphic function in the neighbourhood of 0 in $\mathbb{C}$, such that $f(0) = 0$, $f'(0) = \lambda$, with $|\lambda| \in]0, 1[$. Show that, for all $\alpha \in \mathbb{C}^*$, there exists a function φ holomorphic near 0, such that $\varphi(0) = 0$, $\varphi'(0) = \alpha$, and for all z in a neighbourhood of 0,

$$f(\varphi(z)) = \varphi(\lambda z).$$

(Denote $B = \{\psi \in W(D),\ \psi(0) = \psi'(0) = 0\}$, and study the mapping F with values in B, defined by

$$F(\varepsilon, \psi)(z) = \frac{1}{\varepsilon} f(\varepsilon \alpha z + \varepsilon \psi(z)) - \alpha \lambda z - \psi(\lambda z)$$

for $\varepsilon \in]-\varepsilon_0, \varepsilon_0[$, ψ in a neighbourhood of 0 in B.)

c) Prove the analogue of b) with $|\lambda| > 1$.

Exercise A.2. Let B be the unit ball of $\mathbb{R}^n$. It is proposed to construct a bounded operator $A_0 : C^{\rho-2}(B) \to C^{\rho}(B)$ such that $\Delta A_0 = \mathrm{id}$ ($\rho > 2$, a non-integer).

a) Construction of an elementary solution of Δ. Show that if $n \geq 3$, the formula $\hat{E}(\xi) = -\frac{1}{|\xi|^2}$ defines $E \in \mathcal{S}'(\mathbb{R}^n)$ such that $\Delta E = \delta$.

If $n = 2$ and $z \in \mathbb{C}$ with $\operatorname{Re} z < 2$, we set

$$\hat{E}_z(\xi) = -\frac{1}{|\xi|^z},$$

which defines a holomorphic function with values in $\mathcal{S}'(\mathbb{R}^2)$.

Show that this function extends analytically to $\mathbb{C} \setminus \{2\}$ via the formula $\hat{E}_z = \frac{1}{(z-2)^2}\Delta_\xi \hat{E}_{z-2}$, and deduce that

$$E = \frac{1}{2\mathrm{i}\pi}\int_{|z-2|=1} E_z \mathrm{d}z$$

is a tempered distribution satisfying $\Delta E = \delta$. (Curious readers may calculate E explicitly.) Deduce the existence of an operator $A_1 : \mathcal{E}'(\mathbb{R}^n) \to \mathcal{S}'(\mathbb{R}^n)$ such that $\Delta A_1 = \mathrm{id}$.

b) Let $u \in \mathcal{S}'(\mathbb{R}^n)$. Let $(u_p)_{p>-1}$ denote its Littlewood–Paley decomposition. Show that there exists $\chi \in C_0^\infty(\mathbb{R}^n \setminus \{0\})$ such that

$$\forall p \geq 0,\ u_p = 2^{-2p}\chi(2^{-p}D)(\Delta u_p),$$

and deduce that, if $\rho > 2$,

$$\Delta u \in C^{\rho-2}(\mathbb{R}^n) \text{ implies } u \in C^\rho(\mathbb{R}^n).$$

c) Using an extension operator $C^{\rho-2}(B) \to C^{\rho-2}(\mathbb{R}^n)$, construct A_0.

Exercise B.1. Assume there exists an injective C^∞ mapping $u_1 : M \to \mathbb{R}^d$ (for some d). Then show that if every metric g can be represented by a $u : M \to \mathbb{R}^N$ (for some N), then every metric g can be represented by an *injective* u (look for u in the form (u_1, u_2) and use point iii) of Section B.2.1.

Exercise B.2. a) Show that one can find $\chi : C_0^\infty(\mathbb{R}^n) \to \mathbb{R}$, not depending on $|x|$, such that $\int {}^t\chi'(y)\chi'(y)\mathrm{d}y = \mathrm{id}$.

b) Let $a \in C_0^\infty(\mathbb{R}^n)$, and let $A \in C^\infty(\mathbb{R}^n)$ with A invertible.

Consider the family of mappings $u_{\varepsilon,y} : \mathbb{R}^n \to \mathbb{R}$ defined by

$$u_{\varepsilon,y} = |\det A(y)|^{1/2}\varepsilon^{1-\frac{n}{2}}\chi\left(A(y)\frac{(x-y)}{\varepsilon}\right)a(y).$$

Show that $\int {}^tu'_{\varepsilon,y}u'_{\varepsilon,y}\mathrm{d}y \to [a(x)]^2\,{}^tA(x)A(x)$ (in C_0^∞) when $\varepsilon \to 0$.

c) Using a partition of unity $1 = \sum \psi_j^2$ on M (see Exercise I.6.1), where $\psi_j \in C_0^\infty(\mathcal{U}_j)$ (where $\{\mathcal{U}_j\}$ is a finite covering of M by open sets where local coordinates exists), write an arbitrary metric as $g = \sum \psi_j^2 g$.

Use a) and b) to deduce that g can be approximated by representable metrics.

Exercise B.3. a) Verify that $v : \mathbb{R}^N \longrightarrow \mathbb{R}^{N'}$ defined by $v(x) = \{x_i, x_j x_k\}$, $j \leq k$, is injective and satisfies (2.2.1).

b) Show that if $u : M \longrightarrow \mathbb{R}^d$ is injective, then $v \circ u$ is injective and satisfies (2.2.1).

Exercise C.1. a) Let $\omega \in \mathbb{R}^n$ be such that, for some $k \in \mathbb{Z}^n$, $k \neq 0$, $\langle k, \omega \rangle = 0$. Let f denote the mapping $f : \mathbb{T}^n \longrightarrow \mathbb{T}^1$ induced by $x \mapsto \langle k, x \rangle$. Show that an integral curve of $\frac{\mathrm{d}x}{\mathrm{d}t} = \omega$ on $\mathbb{T}^n$ satisfies, for all t, $f(x(t)) = f(x(0))$. Deduce that the image of x is not dense.

b) Assume, on the contrary (2.1.1) of Section C.2.1: show that then, for any continuous function $f : T^n \longrightarrow \mathbb{R}$ and any trajectory $x(t)$,

$$\lim_{T \to +\infty} \frac{1}{T} \int_0^T f(x(t))\mathrm{d}t = \int_{\mathbb{T}^n} f(x)\mathrm{d}x.$$

(Approximate f by trigonometric polynomials.) Deduce that the image of x is dense, by reasoning by contradiction.

Exercise C.2 $(*)$. Suppose that $\omega \in \mathbb{R}^n$ satisfies

$$(2.1') \qquad \exists \tau > n-1,\ \forall k \in \mathbb{Z}^n,\ k \neq 0,\ |\langle k, \omega \rangle| \geq \frac{C}{|k|^\tau},$$

and take $s > \tau$.

Splitting up the terms of the series according to a dyadic partition of the values of $|k|$ and $|\langle k, \omega \rangle|$, show that

$$\sum_{k \neq 0} |k|^{-s} |\langle k, \omega \rangle|^{-1} \leq \text{Const.} + \sum_{p \geq 0} \sum_{j \geq 0, 2^{p\tau - j} \geq C} 2^{-ps} 2^j N(2^{p+1}, 2^{-j})$$

where $N(r, \varepsilon)$ denotes the number of $k \neq 0$ such that $|k| < r$ and $|\langle k, \omega \rangle| < \varepsilon$.

Show that $N(r, \varepsilon)\varepsilon^{-(n-1)/\tau} \leq C r^{n-1}$ and deduce the convergence of the series.

Exercise C.3. Prove that the composition of two tame mappings is tame.

Exercise C.4. Prove that if a mapping ϕ is tame and linear, then it satisfies

$$|\phi(u)|_s \leq C|u|_{s+r}$$

(with the notation of the book).

The next three exercises establish a tame estimate for the solution of a positive symmetric system in the sense of Exercise II.C.7. Exercises C.5 and C.6 can also treated separately.

Exercise C.5 (*). (A weak and tame Gårding inequality) On $\mathbb{T}^n$, consider the differential operator $A = -\sum_{i,j} \partial_i a_{ij} \partial_j$, where the a_{ij} are C^∞ functions with values in the $N \times N$ Hermitian matrices, which satisfy (for some $\gamma > 0$)

$$\forall x \in \mathbb{T}^n,\ \forall \xi \in \mathbb{R}^n,\ \sum_{i,j} a_{ij}(x)\xi_i\xi_j \geq \gamma|\xi|^2 \mathrm{id}$$

(in the sense of Hermitian matrices).

It is proposed to establish the inequality

$$(*) \qquad \forall u \in C^\infty(\mathbb{T}^n, \mathbb{C}^N)(Au, u) \geq \frac{\gamma}{2}|u|_1^2 - C_0|u|_0^2$$

where C_0 is a constant depending only on γ and on

$$[A]_1 = \sup_{x\in\mathbb{T}^n} \sup_{i,j,k} |\partial_k a_{ij}(x)|.$$

a) Prove the inequality (*) if the a_{ij} are constants (with $C_0 = 0$ and γ instead of $\gamma/2$). Deduce that any point x has an open neighbourhood ω_x whose diameter is controlled using $[A]_1$, such that

$$\forall u \in C_0^\infty(\omega_x, \mathbb{C}^N)(Au, u) \geq \frac{3\gamma}{4}|u|_1^2.$$

b) Using a partition of unity $\sum \varphi_j^2 = 1$ (Exercise I.6.1) adapted to a finite covering $\mathbb{T}^n = \bigcup_j \omega_{x_j}$, prove the inequality (*).

Exercise C.6 (*). (Tame estimates for a commutator) Let $h = h(\xi)$ be a C^∞ function which is homogeneous of degree m for $|\xi| \geq 1$, and let $a \in C_0^\infty(\mathbb{R}^n)$. Then Theorem 4.1 of the symbolic calculus (Chapter I) and the L^2 continuity theorem imply that for all $s \in \mathbb{R}$, $[h(D), a]$ maps $H^{s+m-1}(\mathbb{R}^n)$ to $H^s(\mathbb{R}^n)$ and more precisely that

$$R_a = [h(D), a] - \frac{1}{\mathrm{i}}\sum_{j=1}^n \frac{\partial a}{\partial x_j}\frac{\partial h}{\partial \xi_j}(D)$$

maps $H^{s+m-2}(\mathbb{R}^n)$ to $H^s(\mathbb{R}^n)$. It is proposed to evaluate the norms of these operators as a function of a. For this, we use the notion of the paraproduct introduced in Exercise II.A.5, with the corresponding notation.

a) Show that $[h(D), T_a]u = \sum_p v_p$, with

$$v_p(x) = 2^{mp}\int \rho(y)(S_{p-2}a(x - 2^{-p}y) - S_{p-2}a(x))u_p(x - 2^{-p}y)\mathrm{d}y,$$

$\hat{\rho}(\xi) = \chi(\xi)h(\xi)$ (where $\chi \in C_0^\infty(\mathbb{R}^n \setminus \{0\})$ is equal to 1 on a neighbourhood of the shell $C_1 = \{\frac{1}{4} \leq |\xi| \leq \frac{5}{2}\}$).

Deduce that

$$|[h(D), T_a]u|_s \leq C\|a\|_1|u|_{s+m-1}$$

and

$$\left|\left([h(D),T_a]-\frac{1}{\mathrm{i}}\sum_j T\frac{\partial a}{\partial x_j}\frac{\partial h}{\partial \xi_j}(D)\right)u\right|_s \leq C\|a\|_2|u|_{s+m-2}$$

where

$$\|a\|_k=\sum_{|\alpha|\leq k}\|\partial^\alpha a\|_0.$$

b) Writing

$$\begin{aligned} h(D)(au) &= h(D)T_a u + h(D)T_u a + r_1, \\ ah(D)u &= T_a h(D)u + T_{h(D)u}a + r_2, \end{aligned}$$

show that, if $s+m>0$, $s>0$,

$$|[h(D),a]u|_s \leq C(\|a\|_1|u|_{s+m-1}+\|u\|_0|a|_{s+m}),$$

$$\left|\left([h(D),a]-\frac{1}{\mathrm{i}}\sum_j \frac{\partial a}{\partial x_j}\frac{\partial h}{\partial \xi_j}(D)\right)u\right|_s$$
$$\leq C(\|a\|_2|u|_{s+m-2}+\|u\|_0|a|_{s+m}).$$

Exercise C.7 (*). Let $L=\sum_{j=1}^n A_j\partial_j+B$ be a positive symmetric $N\times N$ system on the torus $\mathbb{T}^n$ (in the sense of Exercise II.C.7).

Suppose that for some $\gamma>0$, $s>0$

(1) $$K(x)=\frac{1}{2}(B+B^*)(x)-\frac{1}{2}\sum_{j=1}^n \partial_j A_j(x)\geq \gamma I,$$

(2) $$\forall\xi\in S^{n-1},\ K(x)+s\sum_{j,k}\partial_j A_k(x)\xi_j\xi_k\geq\gamma I.$$

Using Exercises C.5 and C.6, show, by revisiting Exercise II.C.7, that

$$\forall u\in C^\infty(\mathbb{T}^n,\mathbb{C}^N),\ \frac{\gamma}{2}|u|_s^2\leq \mathrm{Re}\,(Lu,u)_s+C(|A|_s+|B|_s)\|u\|_1^2$$

where C depends only on γ, $\|A\|_2$ and $\|B\|_1$.

Exercise C.8. Let us prove the following result: let $u_\theta\in C^\infty$ $(\theta\geq\theta_0)$ be a family of functions $u_\theta: M\to\mathbb{R}$ such that $\|u_\theta\|_{a_j}\leq M\theta^{b_j-1}$, $j=1,2$, with $b_1<0<b_2$, $a_1<a_2$.

Let us define ν by $\nu b_1+(1-\nu)b_2=0$ and $a=\nu a_1+(1-\nu)a_2$: if $a\notin\mathbb{N}$, then

$$u=\int_0^{+\infty}u_\theta\mathrm{d}\theta\in C^a$$

and

$$\|u\|_a \leq CM.$$

a) Show that we can reduce to the case where $0 \leq a_1 < a_2 \leq 1$.

b) Set $v_0 = \int_{\theta_0}^{\theta} u_{\theta'} \mathrm{d}\theta'$, $w_\theta = \int_{\theta}^{+\infty} u_{\theta'} d\theta'$. Estimate $\|u_\theta\|_{a_2}$ and $\|w_\theta\|_{a_1}$. Then write

$$\begin{aligned} |u(x) - u(y)| &\leq |v_\theta(x) - v_\theta(y)| + |w_\theta(x) - w_\theta(y)| \\ &\leq (x-y)^{a_2} \|v_\theta\|_{a_2} + |x-y|^{a_1} \|w_\theta\|_{a_1} \end{aligned}$$

and optimize the estimate obtained by an appropriate choice of θ.

Exercise C.9. State and verify the Nash–Moser theorem in Sobolev spaces $H^s(M)$.

One might replace Lemma 4.2.1 (Exercise C.8) by the following: let (u_θ) be a family in $C^\infty(M)$ such that $|u_\theta|_{s_j} \leq M(\theta)\theta^{\sigma_j - 1}$ for $j = 1, 2$, $\sigma_1 < 0 < \sigma_2$, and $\int_1^\infty M(\theta)^2 \frac{\mathrm{d}\theta}{\theta} < +\infty$. Then $\int_1^\infty u_\theta \mathrm{d}\theta \in H^s$, with $s = \nu s_1 + (1-\nu)s_2$, $\nu\sigma_1 + (1-\nu)\sigma_2 = 0$.

Remark that the Nash–Moser theorem (C^∞ or H^s) remains true if the hypotheses $(\mathcal{H}_1)$ and $(\mathcal{H}_2)$ only apply for α (or s) in a bounded interval, on which you should state the constraints explicitly.

Exercise C.10 (*). Using Exercises C.7 and C.9, show that the following result holds. Let $F : \mathbb{T}^n \times \mathbb{R}^N \times (\mathbb{R}^N)^N \to \mathbb{R}^N$ be a C^∞ function, $F = F(x, u, p_1, \ldots, p_n)$ satisfying the following conditions:

i) $F(x, 0, 0) = 0$,

ii) $\exists \gamma > 0$,

$$\begin{aligned} K(x) &= \frac{1}{2}\left(\frac{\partial f}{\partial u}(x,0,0) + \left(\frac{\partial F}{\partial u}\right) * (x,0,0)\right) \\ &\quad -\frac{1}{2}\sum_j \frac{\partial^2 F}{\partial x_j \partial p_j}(x,0,0) \geq \gamma I, \end{aligned}$$

$$\forall \xi \in S^{n-1},\ K(x) + s_0 \sum_{j,k} \frac{\partial^2 F}{\partial x_j \partial p_k}(x,0,0)\xi_j\xi_k \geq \gamma I$$

for some $s_0 > \frac{n}{2} + d$.

Then for some s_1, and, for all $s \in]\frac{n}{2} + d, s_1]$, there exists a neighbourhood W of 0 in H^s such that, for all $f \in W$, the equation $F(x, u\frac{\partial u}{\partial x_1}, \ldots, \frac{\partial u}{\partial x_n}) = f$ has a solution $u \in H^s$. (State the constraints on d and on s_1 explicitly.)

Bibliography

Works to be read prior to this book

[CP] J. Chazarain and A. Piriou, *Introduction à la théorie des équations aux dérivées partielles linéaires*, Gauthier-Villars, 1981

[R] W. Rudin, *Functional Analysis*, McGraw-Hill, 1973

[S] L. Schwartz, *Méthodes mathématiques pour les sciences physiques*, Hermann, Paris, 1965

[T1] M. E. Taylor, *Partial Differential Equations. 1: Basic Theory*, Applied Mathematical Sciences 115, Springer-Verlag, Berlin, 1996

[Z] C. Zuily, *Éléments de distributions et d'équations aux dérivées partielles. Cours et problèmes résolus*, Dunod, Paris, 2002

Works accessible at the level of this book

[CM] R. R. Coifman and Y. Meyer, *Au-delà des opérateurs pseudo-différentiels*, Astérisque, vol. 57, SMF, Paris, 1978

[Ha] R. S. Hamilton, *The inverse function theorem of Nash and Moser*, Bull. Amer. Math. Soc. **7** (1982), 65–222

[H1] L. Hörmander, *The analysis of linear partial differential operators*, vol. 1, Springer, 1983

[H2] ——, *On the existence and the regularity of linear pseudo-differential equations*, Enseign. Math. **18** (1971)

[H3] ——, *The analysis of linear partial differential operators*, vol. 3, §18.1, Springer, 1985

[Ma1] A. J. Majda, *Compressible fluid flow and systems of conservation laws in several variables*, Appl. Math. Sci., vol. 53, Springer, 1984

[Mo] J. Moser, *A rapidly convergent iteration method and nonlinear differential equations*, Ann. Sc. Norm. Super. Pisa Cl. Sci. **20** (1966), 499–535

[N] L. Nirenberg, *Lectures on linear partial differential equations*, Regional Conference Series in Mathematics, vol. 17, AMS, Providence, R.I., 1973

[Ra] J. Rauch, *Singularities of solutions to semilinear wave equations*, J. Math. Pures Appl. **58** (1979), 299–308

[Sp] M. Spivak, *Calculus on manifolds. A modern approach to classical theorems of advanced calculus*, Benjamin, 1965

[T] M. E. Taylor, *Pseudodifferential operators*, Princeton, 1981

Complementary works at the research level

[A] S. Alinhac, *Paracomposition et opérateurs paradifférentiels*, Comm. Partial Differential Equations **11** (1986), 87–121

[Au] T. Aubin, *Inégalités d'interpolation*, Bull. Sci. Math. **105** (1981), 229–234

[B1] J.-M. Bony, *Calcul symbolique et propagation des singularités pour les équations aux dérivées partielles non linéaires*, Ann. Sci. École Norm. Sup. (4) **14** (1981), 209–246

[B2] ——, *Second microlocalization and propagation of singularities for semilinear hyperbolic equations*, Workshop and symposium on hyperbolic equations and related topics, Kokota and Kyoto, 1984

[Bt] J.-B. Bost, *Tores invariants des systèmes dynamiques hamiltoniens*, Séminaire Bourbaki 1984-85, no. 639, Astérisque, vol. 133–134

[DJ] G. David and J.-L. Journé, *A boundedness criterion for generalized Calderon–Zygmund operators*, Ann. Math. **120** (1984), 371–397

[G] M. Günther, *On the perturbation problem associated to isometric embeddings of Riemannian manifolds*, Ann. Global Anal. Geom. **7**(1) (1989), 69–77

[H4] L. Hörmander, *Fourier integral operators I*, Acta Math. **127** (1971), 79–183

[H5] ——, *The analysis of linear partial differential operators*, vols. 3 and 4, Springer, 1985

[H6] ——, *The boundary problem of physical geodesy*, Arch. Ration. Mech. Anal. **62** (1976), 1–52

[H7] ——, *On the Nash–Moser implicit function theorem*, Ann. Acad. Sci. Fenn. Math. Series A.I. Mathematics **10** (1985), 255–259

[H8] ——, *Pseudo-differential operators of type 1,1*, Commun. Partial Differential Equations **13** (1988), 1085–1111

[H9] ——, *The Nash–Moser theorem and paradifferential operators*, Analysis, et cetera, pp. 429–449, Academic Press, Boston, MA, 1990

[Hw] I. L. Hwang, *The L^2 boundedness of pseudodifferential operators*, Trans. Amer. Math. Soc. **302** (1987), 55–76

[L] G. Lebeau, *Deuxième microlocalisation sur les sous-variétés isotropes*, Ann. Inst. Fourier (Grenoble) **35** (1985), 145–216

[Ma2] A. J. Majda, *The stability and the existence of multidimensional shock fronts*, Mem. Amer. Math. Soc. **41**, no. 275 and **43**, no. 281 (1983)

[M] Y. Meyer, *Remarque sur un théorème de J.-M. Bony*, Suppl. Rend. Circ. Mat. Palermo (2), Atti del Seminario di Analisi Armonica, Pisa, no. 1, (1981)

[Na] J. Nash, *The imbedding problem for Riemannian manifolds*, Ann. Math. **63** (1956), 20–63

[Sj] J. Sjöstrand, *Singularités analytiques microlocales*, Astérisque, vol. 95, SMF, Paris, 1982

[T2] M. E. Taylor, *Partial Differential Equations. 2: Qualitative Studies of Linear Equations*, Applied Mathematical Sciences 116, Springer-Verlag, New York, 1996

[T3] ——, *Partial Differential Equations. 3: Nonlinear Equations*, Applied Mathematical Sciences 117, Springer-Verlag, Berlin, 1996

Main notation introduced

Index